Cross-border Oil and Gas Pipelines and the Role of the Transit Country

Other Palgrave Pivot titles

G. Douglas Atkins: **T.S. Eliot Materialized: Literal Meaning and Embodied Truth**

Martin Barker: **Live To Your Local Cinema: The Remarkable Rise of Livecasting**

Michael Bennett: **Narrating the Past through Theatre: Four Crucial Texts**

Arthur Asa Berger: **Media, Myth, and Society**

Hamid Dabashi: **Being a Muslim in the World**

David Elliott: **Fukushima: Impacts and Implications**

Milton J. Esman: **The Emerging American Garrison State**

Kelly Forrest: **Moments, Attachment and Formations of Selfhood: Dancing with Now**

Steve Fuller: **Preparing for Life in Humanity 2.0**

Ioannis N. Grigoriadis: **Instilling Religion in Greek and Turkish Nationalism: A "Sacred Synthesis"**

Jonathan Hart: **Textual Imitation: Making and Seeing in Literature**

Akira Iriye: **Global and Transnational History: The Past, Present, and Future**

Mikael Klintman: **Citizen-Consumers and Evolutionary Theory: Reducing Environmental Harm through Our Social Motivation**

Helen Jefferson Lenskyj: **Gender Politics and the Olympic Industry**

Christos Lynteris: **The Spirit of Selflessness in Maoist China: Socialist Medicine and the New Man**

William F. Pinar: **Curriculum Studies in the United States: Present Circumstances, Intellectual Histories**

Henry Rosemont, Jr.: **A Reader's Companion to the Confucian** *Analects*

Kazuhiko Togo (*editor*): **Japan and Reconciliation in Post-war Asia: The Murayama Statement and Its Implications**

Joel Wainwright: **Geopiracy: Oaxaca, Militant Empiricism, and Geographical Thought**

Kath Woodward: **Sporting Times**

palgrave▸pivot

Cross-border Oil and Gas Pipelines and the Role of the Transit Country: Economics, Challenges, and Solutions

Ekpen James Omonbude

palgrave macmillan

DOI: 10.1057/9781137274526

© Ekpen James Omonbude 2013

All rights reserved. No reproduction, copy or transmission of this publication may be made without written permission.

No portion of this publication may be reproduced, copied or transmitted save with written permission or in accordance with the provisions of the Copyright, Designs and Patents Act 1988, or under the terms of any licence permitting limited copying issued by the Copyright Licensing Agency, Saffron House, 6–10 Kirby Street, London EC1N 8TS.

Any person who does any unauthorized act in relation to this publication may be liable to criminal prosecution and civil claims for damages.

The author has asserted his right to be identified as the author of this work in accordance with the Copyright, Designs and Patents Act 1988.

First published 2013 by
PALGRAVE MACMILLAN

Palgrave Macmillan in the UK is an imprint of Macmillan Publishers Limited, registered in England, company number 785998, of Houndmills, Basingstoke, Hampshire RG21 6XS.

Palgrave Macmillan in the US is a division of St Martin's Press LLC, 175 Fifth Avenue, New York, NY 10010.

Palgrave Macmillan is the global academic imprint of the above companies and has companies and representatives throughout the world.

Palgrave® and Macmillan® are registered trademarks in the United States, the United Kingdom, Europe and other countries.

ISBN: 978–1–137–27453–3 EPUB
ISBN: 978–1–137–27452–6 PDF
ISBN: 978–1–137–27451–9 Hardback

A catalogue record for this book is available from the British Library.

A catalog record for this book is available from the Library of Congress.

www.palgrave.com/pivot

DOI: 10.1057/9781137274526

Contents

List of Illustrations		vi
List of Abbreviations		viii
1	Introduction	1
2	The Economics of Cross-border Oil and Gas Pipelines Involving Transit	10
3	The Role of Bargaining in Oil and Gas Transit Pipelines	35
4	Bargaining Positions of the Parties to a Transit Pipeline: Four Case Studies	57
5	The Role of the Energy Charter Treaty: A Critique	101
6	A Case for Mutual Dependencies	117
7	Concluding Remarks	138
References		145
Index		153

List of Illustrations

Figures

1.1	Growth in international oil trade, 1990–2010	2
1.2	Growth in cross-border gas trade, 1991–2010, by transport mode	2
2.1	Transportation costs as a factor of pipeline diameter	12
2.2	Economic rent and profit	24
3.1	Factors influencing the bargaining outcome	51
5.1	Signatories to the Energy Charter Treaty (as of January 2003)	105

Tables

2.1	Oil versus gas pipeline requirements	16
3.1	Similarities and differences in bargaining outcomes for oil and gas	55
4.1	Factors influencing the success or failure of cross-border pipelines	59
4.2	Equity participation in the BTC Company	61
4.3	Equity participation in the South Caucasus Pipeline	63

4.4 Equity participation in the West African Gas Pipeline Company — 64
4.5 Corruption index for selected countries, 2010 — 84

Box

5.1 Definitions of transit in the Energy Charter Treaty (Article 7(10)(a)) — 109

List of Abbreviations

ACG	Azeri–Chirag–Guneshli
BTC	Baku–Tbilisi–Ceyhan
COTCO	Cameroon Oil Transportation Company
EBRD	European Bank for Reconstruction and Development
ECOWAS	Economic Community of West African States
ECT	Energy Charter Treaty
ECTP	Energy Charter Protocol on Energy Transit
EIA	Energy Information Administration
ESMAP	(World Bank) Energy Sector Management Assistance Programme
FDI	foreign direct investment
GATT	General Agreement on Tariffs and Trade
GDP	gross domestic product
IFC	International Finance Corporation
IMF	International Monetary Fund
IPI	Iran–Pakistan–India
LNG	liquefied natural gas
MNC	multinational company
NATO	North Atlantic Treaty Organization
NREP	Northern Route Export Pipeline
PSA	production sharing agreement
SCP	South Caucasus Pipeline
SOCAR	State Oil Company of the Azerbaijan Republic
WAGP	West African gas pipeline
WAPCo	West African Gas Pipeline Company
WREP	Western Route Export Pipeline
WTO	World Trade Organization

1
Introduction

Abstract: *The past two decades have witnessed a significant increase in cross-border trade in oil and gas. It is anticipated that there will be an increase in the number of oil and gas pipelines as a result of the discovery of reserves in remote and land-locked locations and the depletion of reserves close to established markets. A number of problems arise from cross-border oil and gas transportation via pipeline. These problems are more acute in the case of pipelines passing through a transit country. Present and future pipelines face the risk of continuous conflict over legal, economic, and political issues. This book analyses cross-border oil and gas pipelines involving transit countries, with a view to addressing the problem of pipeline disruptions by the transit country. It focuses on the behaviour of the transit country prior to the commencement of operation of the pipeline and on how that behaviour changes after the pipeline has been built and put into operation.*

Omonbude, Ekpen James. *Cross-border Oil and Gas Pipelines and the Role of the Transit Country: Economics, Challenges, and Solutions.* Basingstoke: Palgrave Macmillan, 2013. DOI: 10.1057/9781137274526.

1.1 Introduction

The past two decades have witnessed a significant increase in cross-border trade in oil and gas. Oil exports from the Middle East, for example, grew by approximately 65% between 1991 and 2010, and exports from the Asia-Pacific region (excluding Japan) and the former Soviet Union grew by 175% and 360%, respectively. Figures 1.1 and 1.2 illustrate this growth.

It is anticipated that there will be an increase in the number of oil and gas pipelines as a result of the discovery of reserves in remote and

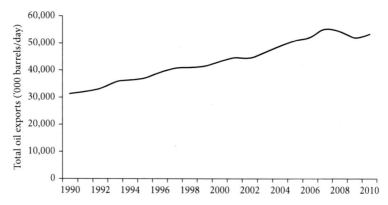

FIGURE 1.1   *Growth in international oil trade, 1990–2010*
Source: Adapted from BP Statistical Review of World Energy 2011.

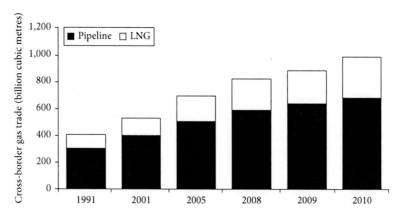

FIGURE 1.2   *Growth in cross-border gas trade, 1991–2010, by transport mode*
Source: BP Statistical Review of Energy 2011.

land-locked locations and the depletion of reserves close to established markets. For example, Azeri oil and gas reserves are far from sea ports or major Western European markets. Statistics show a significant increase in million tonnes of oil equivalent exported from the former Soviet Union in the past decade. This can be argued to demonstrate the increasing role of pipelines in linking markets to the predominantly land-locked reserves of this region. Transporting energy commodities to various locations from such land-locked regions is dependent in some cases on pipelines passing through one or more transit countries. The implementation of successful transit pipeline projects is, therefore, crucial for the security of supply of these commodities. In the case of natural gas, the closest substitute for piping gas is transportation in the form of liquefied natural gas (LNG), which is dependent on coastal access for the LNG shipment. The economics of pipelines compared with the costly LNG process suggest that only at distances in excess of 3,000 miles would the LNG option be more competitive (Avidan, 1997).

A number of problems arise from cross-border oil and gas transportation via pipeline. These problems, which are more acute in the case of pipelines passing through a transit country, fall into three broad categories: reconciling the interests of the different parties involved, the lack of an overarching legal regime to regulate activities, and rent-sharing among the parties (ESMAP, 2003). Specifically, transit oil and gas pipelines face potential disruption by the transit country. Recent developments in the gas dispute between Russia and Ukraine demonstrate the role of transit pipelines in the security of energy supply, as well as the importance of a sufficient understanding of fundamental transit pipeline economics.

Present and future pipelines face the risk of continuous conflict over legal, economic, and political issues. Once the pipeline has been built and put into operation, the risk arises of disruption of the pipeline by the transit country over disputed transit terms. This is due to two key factors: first, bargaining power shifts in favour of the transit country upon construction and operation of the pipeline; second, price changes that result from changes in the value of the throughput can affect the behaviour of the transit country. This is defined as the *obsolescing bargain* – a term coined by Raymond Vernon (1971). In the literature, the obsolescing bargain is a situation in which bargaining power shifts from a multinational company (MNC) to a host country government after investments have been made in a project and the project has started operations (Vernon, 1971). The concept explains the relations between the MNC and the host country as a function of the goals and resources of each party and the constraints it faces. It argues that relative gains are positively related to the relative bargaining power of the parties; that is, the greater one party's bargaining power, the greater that party's share of the gains (Eden *et al.*, 2004). The literature suggests that the party with more resources, fewer constraints, and greater coercive ability gains more from the bargain. According to the original definition of the concept, relative bargaining power at first favours the MNC, which has the initial advantage of being able to invest in several locations – that is, it has alternatives and is, therefore, mobile. The host country offers incentives to attract foreign investment because of the MNC's initial bargaining advantage. According to Eden *et al.* (2004), this bargain then obsolesces over time. Once the investment has been made, the MNC can be held hostage by an opportunistic host country. The longer the MNC is in the host country, and the more profitable the investment is, the more likely it is that the government's perception of the benefit–cost ratio offered by the MNC will worsen. Other factors (e.g. technological spillovers and economic development) encourage the host country to become less resource dependent on the MNC over time. Such behaviour by host governments was evidenced in the wave of nationalisations in developing countries during the 1970s.

Transit oil and gas pipelines are susceptible to the obsolescing bargain because they are characterised by very high fixed costs and relatively low operating (variable) costs (Stevens, 1996). The *bygones rule* states that even a loss-making project will continue to operate for as long as operating costs are met and some fixed costs can be recovered. The implication is that the transit country can continue to increase its demands so long as the pipeline continues to meet its operating costs. In addition, pipelines are inflexible. The cost and security of supply implications of disruptions to an operating transit pipeline are huge, particularly in the case of gas. This enhances the bargaining position of the transit country and tempts it to extract more from the pipeline.

1.2   Objective of the study and main questions

This book analyses cross-border oil and gas pipelines involving transit countries, with a view to addressing the problem of pipeline disruptions by the transit country. It focuses on the behaviour of the transit country prior to the commencement of operation of the pipeline and on how that behaviour changes after the pipeline has been built and put into operation. It also looks at the problems associated with such pre- and post-construction behaviour. Given the nature of the obsolescing bargain as it applies to transit pipelines, and the problem of arbitrary disruption of the pipeline by the transit country as a sovereign state, this research aims to answer two questions. The first is how to define the characteristics of the transit fee in cross-border oil and gas pipeline agreements. The second is how to address the consequences of shifts in power to the transit country. More specifically:

1   Is it possible to have a pipeline agreement that supports the principles of reasonableness, objectiveness, transparency, and non-discrimination such that the transit country will not disrupt the pipeline in the future?
2   How can the shifts in bargaining power among the parties to a cross-border oil and gas pipeline agreement be managed such that the potential for disputes is reduced or removed and security of supply is sustained or enhanced?

This book, therefore, consists of two parts. The first part analyses the economic characteristics of cross-border oil and gas pipelines and the behaviour of the parties prior to construction and after the pipeline has been built and put into operation. The second part uses actual transit pipeline cases to investigate how the consequences of shifts in bargaining power might be mitigated.

1.3   Significance of this book

There are four major reasons why this study is important. The first reason is connected to the obsolescing bargain and the increasing involvement of transit countries in oil and gas pipelines. Compensation to transit countries takes the form either of a transit fee or of an off-take of the commodity, or both (Vinogradov, 2001). A problem arises when transit countries arbitrarily seek to renegotiate transit terms in the pipeline agreement. This leads to disruption of the pipeline, with implications for costs and security of supply. This situation is not helped by the lack of agreement concerning the economic basis for setting the transit fee. Vinogradov (2001), for example, suggests that the transit fee is compensation paid to the transit country for allowing right of way.

The fee, in this view, is a reward to the transit country for sacrificing its sovereignty. The defect of this definition is that provided the transit country signed the agreement willingly, it cannot be argued to have lost its sovereignty. Another view is that the fee is compensation for the negative impact (or externalities) of the pipeline. However, the externalities created by the pipeline can be internalised, and usually the land used for the pipeline is paid for after negotiations between the company and the landowner (ESMAP, 2003; Stevens, 2009). The underlying fact is that there is value attached to the pipeline, and the transit country, by virtue of its contribution to the creation of this value, deserves some share of it. How such contribution to value is created, and how to ensure that the transit country does not damage the pipeline in pursuit of its perceived reward, remain unclear. Shifts in bargaining power to the transit country after the pipeline has been built and put into operation could encourage the transit country to seek to renegotiate transit terms on the basis of its perception of its value to the project. The resolution of this problem is, therefore, important to the success or failure of oil and gas transit pipelines.

Second, there have been attempts to address the problems of cross-border oil and gas pipelines using international instruments or institutions such as the General Agreement on Tariffs and Trade (GATT) and, more specifically, the Energy Charter Treaty (ECT). There is a specific provision in Article 7 of the ECT that addresses the transit of energy. The importance of energy transit is further reflected in the Energy Charter Secretariat's proposed Protocol on Energy Transit (ECTP). The ECT requires that all transit pipeline agreements be characterised by reasonableness, objectiveness, transparency, and non-discrimination. As noted in Chapter 5, these are tenets taken from other legal instruments for international trade (notably GATT). These characteristics are vague at best, simply because of the nature of transit pipelines (as analysed in Chapters 2 and 3). Also, the ECT does not specify the context in which such agreements can be defined as required. The Energy Charter Secretariat appears, therefore, to be struggling with this problem, as there are differing views as to a clear and definitive basis for transit fees.

Third, there is very little literature on the subject of cross-border oil and gas pipelines involving transit countries. Most of the available studies deal with the technical aspects of pipelines (e.g. Masseron, 1990; McLellan, 1992; Mansley, 2003); some deal with the legal aspects (Vinogradov, 2001); and a very few focus on the economic and policy aspects (Stevens,

1996, 2000b; ESMAP, 2003). A few studies have acknowledged that the disruption of the pipeline by transit countries is a bargaining problem (ESMAP, 2003; Hubert and Ikonnikova, 2003; Omonbude, 2007, 2009). It is argued in some of the available literature that the specifics of each pipeline require that cross-border oil and gas pipeline problems be treated on a case-by-case basis. However, fundamental issues that are common to these projects suggest a broad solution to the problem of transit countries disrupting the pipeline as a result of renegotiating the transit terms.

The fourth source of the importance of this study lies in oil price volatility. The implication of volatile oil prices for current and future pipelines is obviously the change in the value of the throughput and, thus, in the rent available from the pipeline. An upward movement in oil prices can be argued to be a determining factor in the aggressive behaviour of the transit country after construction and operation, as is shown in the case of one of the pipelines studied here.

There is little in the literature that is specific to transit country behaviour before and during operation of the pipeline. Previous work has focused primarily on the legal aspects of cross-border oil and gas pipelines, third-party access to gas pipelines, and their legal implications, and a number of publications cover the role of the ECT. These studies have not explored the consequences of transit country behaviour after the pipeline has been built and put into operation. In addition, the general conclusion of these studies is that dispute settlement mechanisms such as those of the ECT are sufficient to deal with such problems case by case. However, this is not applicable here because of the inherent difficulties of binding any sovereign state to anything that includes a settlement mechanism. This study finds factors within and outside the pipeline projects that are common to cross-border oil and gas pipelines and are, therefore, generally applicable.

Some studies have considered the behaviour of the transit country in these negotiations as a bargaining problem. The literature on bargaining in the transit pipeline context includes game-theoretic approaches to explaining the bargaining positions of the various parties to the pipeline (especially gas pipelines, e.g. Hubert and Ikonnikova, 2003; Ikonnikova, 2006). Hubert and Ikonnikova (2003) recognise the shift in bargaining power to the transit country after the pipeline becomes operational, and they propose strategic investment in other pipeline capacity as a possible way to check this shift. Ikonnikova (2006) extends the analysis to describe the game in a partition function form because the Shapley value function applied in earlier studies of the Eurasian gas supply game is prevented from solving the game by externalities (e.g. the possibility of coalitions being formed by other parties, notably the transit country). However, as shown in Chapter 6, this solution is complex and restrictive; that is, it is specific to gas pipelines and situations in which there are actual alternative pipelines to allow for possible coalitions. This book considers the strategic investment solution in a much simpler context. It also identifies a number of other factors that could serve to check the consequences of a shift in bargaining power to the transit country.

This book essentially does three things. First, it analyses the technical and economic aspects of transit oil and gas pipelines to establish how pipelines work and what characteristics make transit pipelines vulnerable to disruption by the transit country. Second, it simplifies the core principles of bargaining and then applies these principles to the oil and gas pipeline context. The premise for this approach is simple: if this is essentially a bargaining problem, it will require a bargaining solution. Thus an explanation of bargaining principles in the context of transit pipelines is crucial to any solution. The previous paragraph acknowledges the literature on game-theoretic bargaining solutions; this approach is focused more on the outcome, but Chapters 3 and 6 demonstrate the importance of applying a framework that combines an outcome approach with a process approach. The approach taken in this book is, therefore, unique in terms of its clear and simplified adaptations of these basic principles to oil and gas pipelines involving transit. Third, the study uses the framework developed from pipeline economics and bargaining principles to analyse actual pipeline cases. It is from this analysis that factors that could mitigate the consequences of a shift in bargaining power to the transit country are defined.

1.4 Structure of the book

This book comprises seven chapters. Chapter 2 first discusses the technical and cost characteristics of pipelines and then applies the concept of economic rent to explain the economics of oil and gas pipelines involving transit.

Chapter 3 investigates the role of bargaining in relation to transit pipelines. It briefly reviews the technical literature on bargaining theory, from which a non-technical set of bargaining principles is derived and applied to transit pipelines in general, guided by the findings from the previous chapter on the economics of pipelines.

The findings from Chapters 2 and 3 form the framework that is applied to the case studies in Chapter 4. Four pipeline projects involving transit have been selected for this research:

- the Baku–Tbilisi–Ceyhan pipeline;
- the Shah Deniz South Caucasus Pipeline;
- the West African gas pipeline project; and
- the Chad–Cameroon pipeline.

This selection has been influenced by the difficulty of obtaining reliable publishable data. Public access to many pipeline project documents is restricted for reasons of commercial confidentiality. The 12 pipeline cases in the UN and World Bank Energy Sector Management Assistance Programme (ESMAP) analysis are classified as being successes or failures on the basis of whether or not they experienced disputes (see Table 4.1). The significant World Bank involvement at some stage during the life of the four pipeline projects selected for this study has encouraged considerable public disclosure of the relevant documents and agreements, which enables more detailed discussion and analysis based both on the criteria set out in the ESMAP paper and on the bargaining principles discussed in Chapter 3. Two oil and two gas pipelines have been selected for this study. The aim of this selection – a secondary objective of this research – is to investigate whether the differences between oil and gas have any effect on the positions of the parties to the transit agreements.

The role of the ECT is examined in Chapter 5 in relation to the pipeline cases discussed in the previous chapter. This chapter compares the findings from the case studies with the provisions of the Treaty (and, especially, the provisions of the proposed Transit Protocol).

In Chapter 6, lessons drawn from Chapters 4 and 5 are developed into a solution to address the consequences of shifts in bargaining powers to the transit country.

Chapter 7 concludes the book.

2
The Economics of Cross-border Oil and Gas Pipelines Involving Transit

Abstract: *This chapter reviews and analyses the literature on the economics of pipelines in general. The primary intention is to identify and define the economic characteristics of cross-border and transit pipelines. The technical, economic, and political factors discussed in this chapter are relevant to an understanding of how pipelines work. The impact of these factors on the transit fee is examined. The aim is to provide a basis for generalisations regarding transit fees, and to provide the basic tools to answer the question of whether a pipeline agreement that is reasonable, objective, transparent, and non-discriminatory does exist or can exist in the long term.*

Omonbude, Ekpen James. *Cross-border Oil and Gas Pipelines and the Role of the Transit Country: Economics, Challenges, and Solutions.* Basingstoke: Palgrave Macmillan, 2013. DOI: 10.1057/9781137274526.

2.1 Introduction

The previous chapter set the scene for this book by showing the importance of cross-border oil and gas pipelines involving transit. The problem of transit countries disrupting the pipeline (through higher transit fees, for example) manifests in conflicts that follow the shift in bargaining power in their favour once a pipeline has been constructed and is in operation. The basis upon which transit fees are determined remains vague and unknown in some cases. The purpose of the transit fee is not explicitly defined in the literature. Some view the transit fee simply as a negotiated compensation or tax paid for right of way, but there are other views as to the purpose of the transit fee, such as:

- compensating the transit state for loss of its sovereignty;
- compensating the transit state for its contribution to the realisation of value added in the actual trade of oil or gas; or
- compensating the transit state for its contribution to the savings created by its position as the most cost-effective route (compared with alternative pipeline routes through other potential transit states).

As argued in section 2.6.1 below, these definitions do not fully explain the purpose of the transit fee in cross-border oil and gas pipeline projects. This raises the question of what would constitute a practicable transit fee (or agreement) – that is, a long-term agreement under which none of the parties (especially the transit country) would arbitrarily seek to renegotiate terms in the future.

This chapter reviews the literature on the economics of pipelines in general. The primary intention is to identify and define the economic characteristics of cross-border and transit pipelines. The technical, economic, and political factors discussed in this chapter are relevant to an understanding of how pipelines work. The impact of these factors on the transit fee is examined in section 2.5. The aim is to provide a basis for generalisations regarding transit fees, and to provide the basic tools to answer the question of whether a pipeline agreement that is reasonable, objective, transparent, and non-discriminatory does exist or can exist in the long term.

2.2 Economic characteristics of pipelines

Pipelines are generally characterised by high fixed costs, as a result of huge capital investments, and relatively low operating costs (compared with the fixed costs). Moreover, large economies of scale prevail. The higher the throughput of the pipeline, which depends on its diameter, the higher the variable costs per unit of commodity (oil or gas in this case) (McLellan, 1992). Also, with continued throughput, much of the initial investment is recovered, leaving less to expropriate. It is important for the pipeline to operate at maximum possible throughput so as to spread the fixed costs over higher output levels over time.

Unit transport costs for the same level of utilisation are lower for pipelines with larger diameters. This is due to the fact that a pipeline's capacity is approximately proportional to two-and-a-half times the square of its diameter, while the capital cost is directly proportional to the diameter of the pipeline.

Figure 2.1 illustrates how fixed and variable costs per unit change with throughput as a factor of pipeline diameter. Of note is the point at which the total unit cost of transportation is at a minimum (the optimum throughput). At this point, most of the capital expenditure has been recovered (hence the continuous decline in fixed costs) and operating costs (which are minor compared with fixed costs) become the major component of the cost outlay for the rest of the life of the project. The

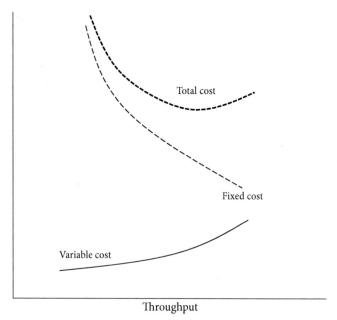

FIGURE 2.1   *Transportation costs as a factor of pipeline diameter*

figure also indicates how low the variable costs of operating a pipeline system are compared with the fixed costs.

2.2.1 Cost characteristics and concepts

Typically, any producer will competitively minimise costs and maximise profits subject to technological constraints. In competitive markets, producers consider input prices during a given period of time to be fixed irrespective of their demand for those inputs (Besanko and Braeutigam, 2002) or their expected output for the period (Layard and Walters, 1978). The same principle applies to pipelines, especially during the construction phase. The objective is usually to minimise the operating costs, as the fixed (capital) costs are deemed to be sunk costs (this is explained further by the bygones rule, which is discussed later in this chapter).

By the nature of pipelines, both fixed and variable costs arise, but they differ from project to project. A typical cost function for pipelines would take the following form:

$$C = F(D) + G(Q, D),$$

where $F(D)$ represents the fixed cost associated with the installation of a pipeline of diameter D, and $G(Q, D)$ are the variable costs (Brito and Rosellon, 2001).

The capital costs of a pipeline have a direct dependence on the pipeline's diameter and length (McLellan, 1992). This can be argued to be the basis upon which some generalisations or theories regarding pipeline economic issues such as rent and transit tariffs have been formulated. (For example, Masseron, 1990, defines a pipeline transport tariff as a cost function tied to the length of the pipeline.) These capital costs are influenced by a number of factors, including:

- mobilisation (or demobilisation) costs of contractors;
- communications and control systems;
- level of difficulty of road and river crossings, as well as terrain;
- way leave costs;
- costs of compressor stations and terminals;
- costs of steel and welding; and
- environmental costs.

Pipeline costs necessarily rise with diameter, since more steel is needed. A further cost effect is that the thickness of the pipe wall needs to be greater as the diameter increases. This, in turn, increases welding costs. This further emphasises the importance of pipelines running at the highest possible capacity, in order to reduce average fixed costs and hasten investment recovery. For oil pipelines, throughput can be enhanced by the application of a *drag-reducing agent* (or 'flow improver') – a chemical that allows the crude oil to flow more easily. This raises the operating capacity of the pipeline. This enhancement, however, is probably subject to decreasing marginal profitability (Pipeline Industries Guild, 1984).

Pipeline operating costs (e.g. engineering, supervision, administration, overheads, interest, and contingencies) tend to be a small fraction of the capital costs over the life of the pipeline. Some of the literature suggests these costs are no more than 4–7% of capital costs (McLellan, 1992, 4%; Brito and Rosellon, 2001, 5–7%). For oil and gas pipelines, operating costs (e.g. the costs associated with pumping oil or compressing gas) will vary with throughput.

Pipeline economic feasibility is also obviously affected by the standard issues of market conditions, geology, technology, and political climate.

An understanding of the relationship between these forces and pipeline costs also shows their relevance to and impact on transit fees. These factors affect both the capital and operating costs of the pipeline, which in turn have implications for the justification and determination of transit fees.

2.2.2 Technical factors of pipeline economic feasibility

From the initial design stage through to the construction and operation of the pipeline, costs necessarily play a significant role in pipeline economics. It is important to take into account both the capital and operating/maintenance costs when choosing the optimum design. It is necessary to perform economic calculations to compare the proposed design with other combinations of pipeline size, operating pressure, and horsepower to ensure the best choice of system (Kennedy, 1993). The design of pipelines involves an attempt to achieve economic balance among the capital cost of the pipeline and its associated pumping or compression facilities and the subsequent annual cost of the pipeline's operation (Pipeline Industries Guild, 1984). The most significant operating cost is the cost of energy required to push the fluid or gas through the pipeline. One major technical factor that influences the cost of pipelines is the load factor; other factors include technical differences between oil and gas, the size of the pipeline, operating structure, compressor costs, and station spacing.

Load factor

Load factor is the percentage ratio of the daily throughput averaged over a year to the maximum throughput on any day in that year (McLellan, 1992):

$$\text{Load factor} = \frac{\text{Average daily throughput over a year}}{\text{Maximum throughput on any day in the same year}} \times 100.$$

This definition is frequently applied to a specific stream of product passing through the pipeline. A pipeline can carry two or more streams, which are likely to be owned by different companies, in which case each stream has its own load factor. The terms 'load factor' and 'utilisation factor' are used interchangeably; however, the utilisation factor is slightly differently defined in some of the literature as the ratio of average throughput of the two or more streams over a year to the maximum daily throughput capacity of the pipeline as built. In other words, the utilisation factor can be defined as a load factor over the streams of commodities passing through the pipeline. There is an inverse relationship between the unit cost of transport in a pipeline system and the utilisation factor. This means that the higher the load factor (or the more complete the utilisation of capital facilities involving predominantly fixed costs), the lower the unit costs (and the faster the recovery of the investment).

Comparison of oil and gas pipelines

There are fundamental differences between oil and natural gas that influence the cost of transporting these hydrocarbons via pipeline. For the purposes of this book, it is important to clarify these differences because they have a significant effect on the economics of pipelines. The key distinction between oil and natural gas pipeline transportation lies in the energy required for pumping (or compressing), and in some cases the diameter of the pipeline. For natural gas, compared with oil, more energy is required for pumping for a given pipeline diameter, and pipelines sometimes require larger diameters, thus making gas pipelining comparatively less economic than oil pipelining. For the same pipeline diameter, length, and load factor, there is more energy content in the amount of oil than in the amount of gas that can be transported (the energy content in one barrel of oil is equivalent to the energy content of 6,000 cubic feet of natural gas). This obviously affects the costs of pumping the hydrocarbons, and the literature suggests that natural gas pumping costs are about four times more than crude oil pumping costs.

TABLE 2.1 Oil versus gas pipeline requirements

	Fuel oil	Natural gas
Line diameter	20 inches	20 inches
Optimum throughput at 100% utilisation rate	160,000 bbl/day	220 MMscfd
Energy content	6,000,000 BTU/bbl	1,000 BTU/scf
Energy throughput (10^9 BTU/day)	970	220
Energy transportation cost per million BTU/mile (proportional)	1	4.5

Table 2.1 summarises the fundamental differences in technical requirements for oil and gas pipelining.

The impact on costs is evident. For example, compression costs are higher for gas than for oil. The brake horsepower required to pump a throughput of oil through a pipeline is constant no matter the level of throughput required. For gas, however, more brake horsepower is required to increase gas pressure; in addition to being directly proportional to the rate of throughput, the brake horsepower required to pump gas is a function of the compression ratio – the ratio between compressor inlet and outlet pressure. Increased throughput requires an increased compression ratio, and thus more power (Yedidiah, 1980; McLellan, 1992). Transmission costs are also higher for gas than for oil, mainly because of the compression cost component:

$$C_T = (C_p + C_{comp} + C_{op} + C_{main} + C_{fuel})/V_T.$$

The main components of gas transmission costs (C_T) are the annual charges related to the pipeline (C_p), the annual charges related to the compressor stations (C_{comp}), operating costs (C_{op}), maintenance costs (C_{main}), and fuel costs (C_{fuel}), all expressed as a fraction of the total annual volume delivered (V_T) (Pipeline Industries Guild, 1984). The main implication is that for gas pipelines to be more cost-effective, they have to operate at high pressure.

Pipeline size

The question of how big a pipeline should be is motivated by the pipeline fundamentals already discussed. According to McLellan (1992), the option of cheap incremental capacity is usually used (e.g. the addition of pumping or compression stations along the pipeline to increase capacity).

However, if this increased capacity is never needed, the result will be a low utilisation factor, thus increasing the unit cost over the life of the pipeline project.

As a factor in the sizing decision, incremental capacity is useful in tariffing the oil or gas of third parties. Once the potential third-party users have been identified (along with their capacities and utilisation factors), the size of the pipeline (incremental capacity) can be determined, which aids in the tariff determination process.

The key factors to consider in the choice of pipeline size include the following:

- an acceptable level of confidence that additional volumes of the oil or gas to be transported will appear, and a level of confidence/satisfaction with their profiles and load factors;
- the features of the fiscal regime relating to the pipeline and its various third-party users;
- the strategy of the investors (e.g. whether the investors' objective is to ensure control of transportation in the area); and
- the policy standpoint of the relevant authorities about such issues as the proliferation of pipelines and the addition of pumping or compression stations, and the authorities' environmental objectives.

It should be noted, however, that in a monopoly firms will – if the need arises – keep output low (undersize) and prices high, which in high-fixed-cost markets constitutes a barrier to entry (Brennan, 2002). This must taken into consideration in policy-making.

Other technical factors

Other factors that have an impact on the economics of oil and gas pipelines include the cost of ensuring the oil or gas transported meets the entry specifications of the pipeline system, the costs of reducing the quality of oil or gas (and thus the value) by varying the composition of the fluid(s) in the pipeline system (e.g. by addition of a drag-reducing agent, as previously mentioned), and the costs of third-party connections to existing pipeline systems. These costs can be quite significant, and they tend to constitute a major component of the overall costs of transporting oil and gas via pipeline. Other factors include determining the number of shippers the pipeline will serve and the number of entry and delivery points for each shipper (as well as an acceptable combination of entry and delivery points).

2.2.3 Pipelines as natural monopolies

Pipelines are generally very capital intensive, and one of the results of their large technical economies of scale is that pipelines possess the characteristics of natural monopolies (Stevens, 1996).

A natural monopoly occurs in a market structure in which a single seller can, as a result of economies of scale, supply the socially optimal quantity of output at the lowest possible total cost. It is *natural* because the economies of scale make it more efficient for one firm to supply a particular good or service to the market than for two or more firms in competition to do so. There may or may not be more than one potential seller ready to take the place of this monopolist (Sharkey, 1982). In a natural monopoly, this monopolist is able to supply the good or service to the entire market at a lower or more efficient cost than the two or more potential suppliers. This is based on the concept of *subadditivity*, defined in the economics literature as follows (Sharkey, 1982; Foster, 1992; Martin, 1994):

> If $q^1, q^2, \ldots, q^k$ represent the output bundles that add up to q, and if the cost of producing the sum of the commodity bundles q^1 to q^k is greater than the physical and organisational costs of a single firm producing q, then a single firm is superior on grounds of efficiency to two or more competing firms, and can jointly produce bundles of q^1 to q^k more cheaply than if they were produced separately or by two or more competing firms.

If there are no potential (inactive) sellers, the monopolist makes price and output choices to maximise profit. The condition of natural monopoly can apply in the case of services provided to a large number of users. The firm has to incur huge costs in the construction of delivery infrastructure (as applies to pipelines), and once this infrastructure is in place, additional users can be accommodated at low marginal cost. This can be argued to be a barrier to entry, because new entrants will have to incur huge sunk costs (costs already incurred by the existing supplier), which again applies (in most cases) to pipelines.

The implication for oil and gas pipelines of the characteristics of natural monopolies is quite straightforward. It is preferable to have one pipeline between two points than to have two or more (or a network of) pipelines of similar capacity competing for the same throughput volumes. According to Stevens (1996), it amounts to economic inefficiency for more than one pipeline to run between two points because the additional pipeline loses potential cost savings from economies of scale. In addition, it is difficult for alternative forms of transport to compete with the existing pipeline. The main alternative forms of transport to pipelines for oil and gas are tankers and, for gas, LNG. As mentioned in Chapter 1, it is not cost efficient to transport LNG by tanker over distances of less than 3,000 miles. Moreover, alternatives to pipeline transport cannot compete with land-locked pipelines. It is, therefore, not unusual for a pipeline to be constructed to be big enough to accommodate all throughput requirements, but this has its own drawback: the exposure of this single pipeline to risk. This is especially the case in situations involving transit through one or more countries, when the bargaining power tends to shift to the transit country (or countries) once the pipeline has been constructed and is operational (the obsolescing bargain concept).

2.3 Economics of cross-border oil and gas pipelines

In the transportation of oil and gas via pipeline across national borders, the pipeline economics are influenced by factors other than (but in addition to) the basic economics discussed in the preceding sections. These factors include regulation, market size and type, and the sharing of benefits.

2.3.1 Regulatory aspects of cross-border pipelines

When a pipeline has to cross the border of another country, state, or territory to reach its market, a major factor in its construction and operation is the problem of different legal and regulatory regimes. This factor affects the successful operation of the pipeline in the following ways (ESMAP, 2003):

- The differences in the legal and regulatory structures of the countries involved, and the attempts by both parties to preserve or uphold their respective regimes because of particular interests, will cause conflict. The lack of an overarching regulator for the whole project further reduces the scope for quick conflict resolution.
- The pipeline project may conflict or compete with a national project. To protect its national project, the country concerned will seek to apply its legal and regulatory powers to the part of the pipeline that crosses its territory or jurisdiction in a way that favours its own project, obviously leading to conflict.

According to Vinogradov (2001), there are two types (or models) of cross-border pipeline agreement. In the first system – the connected national pipelines model – each national section of the pipeline is under the territorial jurisdiction of the respective country and is, therefore, subject to that country's domestic law. Regulation in this case takes the form of contracts between the owners or operators of the pipelines or agreements between the respective governments, or a mixture of both. The benefit of this sort of arrangement is that the parties are able (to an extent) to protect their respective interests, depending on the ownership structure of the pipeline project. The drawback lies in the increased political risk and uncertainty, as well as the apparent lack of limits to the lengths to which the respective parties can go in protecting their interests.

The second system involves an international pipeline project that is designed and operated as a single integrated unit. Regulation in this regard involves a system of interrelated agreements (both inter-governmental and commercial), and the pipeline project is protected by the inter-governmental agreements reached between the parties involved. There still exists the potential for conflict in arriving at acceptable terms of agreement between the parties, in addition to the possibility of disagreements regarding how benefits are shared, before and during the operation of the pipeline.

2.3.2   The types and sizes of market involved

The structures of the respective national energy markets may differ in terms of regulation, the level of competition, or even demand for the commodity. In the case of gas pipeline projects, for example, it is important to recognise the differences between the markets involved. Each market could be either a commodity supply market, and thus subject mainly to market forces, or a project supply market with different objectives, such as general economic development. The pipeline project must operate under such contrasting situations. Reconciling these different interests within an overarching legislative or regulatory structure is difficult, as the parties involved will seek to protect their own interests, which will lead to conflict.

2.3.3   Benefit-sharing in cross-border pipelines

Cross-border pipeline projects attract profits or rent, which must be shared across the border. The basis for sharing these profits has to be clearly defined and acceptable to both parties. This is a potential source of conflict simply because the parties involved – in protecting their interests – are likely to differ on the method of and justification for sharing these benefits. This is a common source of dispute in the determination of transit fees, and a more detailed discussion follows in section 2.5.

2.3.4 Political/diplomatic relations

Border disputes can affect the successful operation of cross-border pipeline projects. The disruption resulting from such conflicts shows in delayed delivery to the market or a complete halt in the transportation process. This has an obvious impact on the cost of the project, as well as security of supply implications for the destination market. Diplomatic relations between the countries involved also have an impact on the economics of cross-border pipelines. For example, it was argued in 1998 that some members of the Russian foreign ministry were not keen on exports from Kazakhstan or Azerbaijan (Soligo and Jaffe, 1998), since Caspian exports serve as competition to Russian local and international production. One could, therefore, argue from a Russian nationalist perspective in favour of preserving the country's role in its 'geographic periphery' (Soligo and Jaffe, 1998). Soligo and Jaffe further argue that deliberate attempts could be made to create obstacles to the resolution of pipeline disputes as a bargaining strategy that worked to Russia's advantage.

2.3.5 Environmental policy

Another factor that affects the economics of cross-border oil and gas pipelines is state policy on the environment, which can play a significant role in their construction and operation. The parties concerned will have to agree on issues ranging from technical (e.g. the size of the pipeline to be constructed, whether the pipeline runs above- or underground, gas emissions) to purely environmental concerns (e.g. the effect on rare and endangered species, the fragmentation of habitats). The implication of environmental policy for the economics of the pipeline becomes obvious in compensation negotiations, which can affect the cost of operating the pipeline.

2.4 Implications of the economic characteristics of cross-border oil and gas pipelines

It is reasonable to suggest that the factors discussed above are difficult to separate. This is because they are interwoven; one leads to (or has an effect on) another. Regulation, for example, comes into play in the nature and structure of the markets involved, and can be strongly influenced by the political or diplomatic dispositions of the countries involved. What draws these factors together is the effect they have on the economics of pipelines in general.

These factors all have cost implications, for example. Conflicts arising from any of these factors are likely to have an effect on the operating costs of the pipeline (assuming the pipeline has started pumping). Taking the bygones rule into consideration, caution must be exercised that these operating costs do not rise to levels that could influence the firm's decision to stay in business.

Another example is the load factor. As already discussed, a pipeline can carry two or more product streams, implying the involvement of more than one firm or government. Each company using the same line will have its own utilisation factor. The higher the utilisation factor, the lower the unit cost of transporting the commodity via the pipeline. Therefore, if one party, for one reason or another (e.g. protection of a national project, low demand for the commodity, or regulatory restrictions) does not fully utilise these capital facilities, this reduces the overall utilisation factor of the pipeline, which in turn has obvious cost implications.

The discussion so far has not linked the economics of pipelines to transit country behaviour post-construction or to the transit fee. It has, however, provided the basis for the discussion in the following section of the economics of transit, the concept of economic rent, and its implication for the transit fee.

2.5 The economics of transit

This discussion of the economics of pipelines has implications for cross-border pipelines and also transit pipelines. The key factor in discussing the implications for transit pipelines is the addition of another player – the transit country – to the cross-border economics. The transit country will expect some 'reward' for its part in the entire trade process, either in the form of an off-take agreement or in the form of a transit fee (which is the main focus of this section, and indeed of this book), for a number of reasons (which are discussed later in this section) ranging from compensation for loss of sovereignty to return for its contribution to value added in the trade process. This section discusses the economics of the transit factor in cross-border oil and gas pipelines, and attempts to justify the transit fee charged by transit country governments in most cross-border pipeline contracts. The concept of economic rent is introduced to help justify the transit fee, as well as to serve as the basis for answering the question of whether it is possible in the long term to have a pipeline with reduced risk of disruption by the transit country.

2.5.1 The concepts of economic rent and transfer earnings

The theory of the pricing of factors in fixed supply is based on the concept of *economic rent*. Economic rent is defined as the payment to a factor over and above what is needed to keep that factor in its current use. According to Koutsoyiannis (1979), it is a payment to this fixed factor in excess of its opportunity cost. If there is no alternative use for the factor, then its opportunity cost is zero; therefore, all its payment is rent.

The concept of economic rent was originally based on rent payment to a single factor – land. The concept, however, extends to all factors of production; therefore, any factor may receive an economic rent (Keirstead and Coore, 1946; Worcester, 1946). A typical example of payment to a factor whose supply is completely inelastic would bear the following feature:

▸ The available quantity of the factor in supply remains fixed irrespective of its price (i.e. an increase or decrease in price cannot induce a corresponding increase or decrease in supply of this factor).

The supply of the fixed factor remains constant. Given an increase in demand and perfectly fixed supply, the price of the factor increases also. The difference between prices is rent. This change in rent paid for the factor is entirely a function of the change in demand for it; therefore, rent is price-determined. It is not a cost in the establishment of the price of the factor. This type of rent is often referred to as *pure economic rent* (Koutsoyiannis, 1979).

As already mentioned, if the fixed factor has no alternative use (or is perfectly inelastic), all its payment is rent. Its opportunity cost is zero, which implies that nothing has to be paid to keep the factor in its current use; its current use is not affected by whether its price is high or low. It can, therefore, be argued that reducing (by way of tax, for example) this payment to a fixed factor that has only one use will not affect its availability/supply. If, however, the factor has some elasticity of supply, only a portion of its price becomes rent. This is illustrated in Figure 2.2.

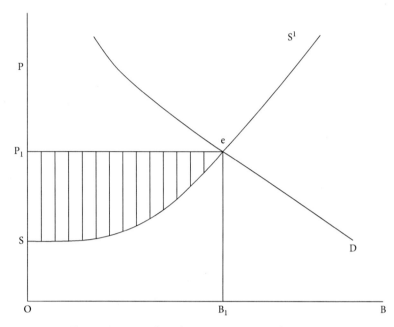

FIGURE 2.2 *Economic rent and profit*

From the figure, the equilibrium price is P^1 and the equilibrium quantity is B_1. The total payment to this factor is represented in the figure by $O\ P^1\ e\ B_1$. The payment can be split into two parts:

- the amount that is paid to keep the factor in its current use; and
- the opportunity cost of the factor.

If the supply curve SS^1 depicts the marginal cost of an additional unit of the factor, then the amount of payment required to keep this factor in its current use is the area $O\ S\ e\ B_1$ on the graph. Therefore, the rest of the total payment to this factor (i.e. $S\ e\ P^1$) is economic rent, as it is in excess of what is needed to keep this factor in its current use. This economic rent is also commonly referred to in the literature as *producer surplus* (i.e. a surplus return to the factor over and above its opportunity cost; Currie et al., 1971). Any factor with less than perfectly elastic supply will earn economic rent. The steeper the supply curve of the factor, the greater its economic rent. If the supply curve were perfectly elastic, the payment would contain no rent.

In some of the literature, the area of the graph ($O\ S\ e\ B_1$) in Figure 2.2 depicting the amount of payment required to keep the factor in its current use is also referred to as *transfer earnings* (Koutsoyiannis, 1979). In labour economics, this is the minimum wage at which a worker is prepared to supply his labour. The worker will continue to work as long as his wage is equal to, or higher than, this minimum payment. A simple example is a football player who earns, say, £500,000 a year for his skill in the game. He earns a lot more than he would in an alternative employment such as working as a window cleaner for, say, £10,000 a year. If he is willing to play football for £10,000 or more, his earnings include £490,000 in rent, and £10,000 would be his transfer earnings. According to Tilton (2004), this rent (i.e. the £490,000 in the illustration) could be taken away without affecting his behaviour or altering the performance of the economy.

Some factors are fixed in the short run and become variable in the long run. *Quasi-rents* are payments to factors of which there is a fixed supply in the short run. Quasi-rents are a residual payment. Fixed inputs cannot be withdrawn from their present use in the short run, while variable inputs can move to alternative uses where there are higher returns. Therefore, these variable inputs are paid their opportunity cost, and what is left (residual payment) is paid to the fixed input. Quasi-rent is thus the difference between total revenue (TR) and total variable costs (TVC). Mathematically, this can be represented as:

Quasi-rent = TR − TVC.

Quasi-rent has been argued to be divisible into two parts (Koutsoyiannis, 1979; Tilton, 2004): total fixed cost (or the opportunity cost of capital and other fixed factors) and excess (or pure) profit. Excess profit is, therefore, defined as the difference between quasi-rent and total fixed cost (TFC):

Quasi-rent = Excess profit + TFC,

or

Excess profit = Quasi-rent − TFC.

It is important to reiterate that quasi-rent disappears in the long run, as all factors become variable in the long run. Payment to these variable factors is enough to keep them in employment or use and earn normal profit, thus removing quasi-rents.

2.5.2 Economic rent and the transit fee

As mentioned at the beginning of this chapter, there are divergent and vague views as to the key basis of, or justification for, the transit fee. One such view is that the transit fee is a payment to the transit country for allowing right of way. Another view defines the transit fee as compensation paid to the transit country for surrendering part of its sovereignty. The problem with these definitions is that the pipeline is still subject to the jurisdiction of this transit country during its construction and operation. Another view regards the fee as a reward for helping to realise the value added in cross-border trade in oil and gas. Yet another view suggests the transit fee is a payment to the transit country for its contribution to the cost savings created by its position as the cheapest route (compared with any alternative routes available). The transit fee is also argued by some to be assessed on the basis of international norms that employ charges per unit of volume per kilometre.[1] These definitions lack clarity, and this lack of clarity can be argued to have played a role in the various disputes over transit fees. These grounds for criticism are as follows (in no specific order):

- The argument restricting the transit fee to a function of the length of the pipeline is straightforward and practicable, but it fails to acknowledge the role of bargaining in cross-border transit trade. The bargaining process could go a long way in influencing the proportion of rent that the transit country can extract from the project. This argument also fails to recognise the role of the transit country. The project would either be of no value without it or be more expensive through another route.
- These definitions fail to acknowledge the inevitability of a transit fee. In reality, the transit country does not require justification for charging a fee; it does so simply because it can!
- As discussed later in this chapter, it is difficult to attach value (for empirical purposes, for example) to these views on the transit fee, owing to their definitional ambiguity (and, arguably, their theoretical invalidity). They do not take into account, for example, what factors should be considered in determining the transit country's actual contribution to the value added in the oil and gas trade (e.g. costs, risks, for both the transit country and the producer country).

Given this ambiguity, this book suggests that the concept of economic rent has significant implications for the transit fee. It can be used to explain the purpose of, justification for, and character of transit fees in oil and gas pipeline trade. We can summarise the review in the previous section in two key points:

- Economic rent is payment to a factor over and above what is required to keep the factor in its current employment in the long run.
- Transfer earnings are payments to a factor sufficient to keep it in its current employment.

Will a cross-border oil or gas pipeline project earn economic rent? The economics of pipelines suggest negligible variable costs compared with the fixed costs. Transport costs are much lower for the same level of utilisation with larger pipeline diameters. The bygones rule shows that even under loss-making conditions, a pipeline project (or any project) will continue to operate so long as operating (variable) costs are met. The possibility of a surplus or rent in this case is enhanced by the industry's perception of capital costs in their books as sunk costs; therefore, their key objective is to minimise the operating costs through increased efficiency to increase profits, and obviously create rent. This answers the question of whether cross-border oil and gas pipeline projects can earn economic rent.

The earlier discussion of the characterisation of economic rent in the literature indicates that taxing this rent (or taking it away) ideally should not alter behaviour; therefore, in an ideal situation, if a transit country seeks to capture all or some of this rent by way of a transit fee, it should not generate dispute, since the project will continue to operate so long as it meets its variable (operating) costs and there is some allowance for the recovery of fixed costs (the bygones rule). This has not been the case in reality, as some pipeline projects have sparked disputes surrounding how this rent is shared among the parties involved. One reason for this is the lack of a clear understanding (or at least an agreed definition) among the parties as to what this rent is, or where to draw the line between profit and supernormal profit, or how big the rent is. An attempt is made in this section to delineate the perceived rent-determining factors in oil and gas pipeline projects.

In trying to establish the rent from a cross-border oil or gas pipeline project, there are two key components to consider: the value of the project and the cost of the project. The interaction between these factors produces rent; that is, the economic rent from the pipeline project can be defined as the difference between the cost of the project and the value of the project.

The value of the project

Most definitions of the value of a project tend to centre on the same general theme: the total worth of the project. In the oil and gas industry, a straightforward way to determine the value of a project would be to define the projected volume of oil or gas to be traded in terms of price (i.e. total revenue, or expected revenue: quantity produced multiplied by price), taking into account a number of factors. The key factors that will determine the value of the project include the following (in no specific order):[2]

- initial investment outlay;
- reserve size;
- production outlay;
- oil/gas prices;
- revenue streams;
- estimated lifespan of the project; and
- geography, geology, and environmental aspects.[3]

Except in the case of projects that are entirely politically motivated, all market-driven project analysis is based on the present value of the difference between the projected (or future) net cash flows of the project and the initial investment (in other words, the net present value). The project will not be started if the net present value figures are not positive. In the literature, the determination of the value of a project is not particularly complicated; most of the determining factors in this case are usually given, and modelled projections thereof (depending on the qualifying assumptions employed in the exercise) can be quite straightforward.

The key to explaining the complexities surrounding the rent that can be generated from a pipeline project lies in the cost component. The factors that determine the cost of the project along the entire value chain are less straightforward than the factors that determine its value. A detailed discussion of these cost factors is justified, therefore, as it might help clarify what the rent actually is.

The cost of the project

As already mentioned, the rent from a project can be argued to be the difference between project value and project cost, but what is this cost?

Again already discussed is that this cost entails the cost profile along the entire value chain of the project. For the purpose of this analysis, the nature of project cost is discussed under three sub-headings:

- the general costs element;
- the profit element; and
- the transit element (or the role of the transit country).

The main objective is to ascertain the factors in the determination of the cost component, to explain how straightforward (or otherwise) it is to obtain and capture rent in oil and gas transit pipeline projects. The three elements are thus discussed in this context.

The general costs element

The factors determining general pipeline costs have been discussed already in this chapter: high fixed costs (initial investments), relatively negligible operating costs (the most significant element being the cost of pumping the oil or gas), optimum throughput, technical aspects, geology, and so on (see section 2.2). This covers only the cost of transporting the commodity to the market. The total sum will also include the costs of exploration and production, which are relevant to the determination of the cost of the project in terms of the entire value chain. The costs involved here are mainly tangible (drilling, pipeline, pump stations, wages, etc.); therefore, this aspect of the determination of the project cost profile can be argued to be straightforward.[4] On this basis alone, determining the rent available from a project would be straightforward: the costs would be summed and deducted from the projected value of the project, and the balance would be available as rent. The problem is, however, more complex, as is shown below. With respect to factors with fairly elastic supply, the concept of economic rent also takes into consideration a profit element. The issue of rent arises only after the factor has met its operating costs and earned enough profit to stay employed.

The profit element

In the development of fiscal systems for oil and gas ventures, consideration is given for the project (or contractor/company) to make a profit (Johnston, 1994). The project must be able to meet its costs and make a return on investment sufficient for it to remain operational. Governments make an allowance for this. The general structure of the costs and allowances for profits, and their relationship to rent, is as follows (Johnston, 1994):

- gross revenues (or the value of the project);
- total profit and a cost recovery element/allowance (which should sum to the project value);
- total cost from the perspective of the government, which comprises exploration, development, and operating costs, as well as the contractor's take (a profit); and
- the balance (taking away the above costs and profit), which is rent and which the government can seek to capture.

In Johnston's analysis, the manner in which governments try to capture this rent takes the form of some combination of bonuses, royalties, production-sharing, taxes on profits, and government participation. This raises the question of how straightforward it is to determine and extract this rent. Of key relevance here is the rate of return on capital used in the calculation of the project's profit. Contractors and companies usually employ different rates than governments in the calculation of profit figures for the project; therefore, it is likely that the total costs from the government's perspective will be lower than the costs from the company's perspective (or vice versa), which leads to a potential for dispute as to what the actual rent is.

It should be noted at this point that the above discussion has focused on the determination and sharing of economic rent, assuming away the transit country element. I believe that the divergence of views about cost can be resolved relatively straightforwardly so long as the project is not complicated by the transit element. This is because for typical exploration and production projects in which the commodity is either consumed by the producer country or shipped directly to the market, all rent is shared between the E and P company and the government or captured entirely by the government. The introduction of the transit country element into this method of determining value and cost reduces the straightforwardness with which rent is obtained and captured.

The transit element

The role of the transit country in cross-border oil and gas pipeline projects is very important, particularly in terms of how the rent from the project is determined and shared. As stated in the previous section, the involvement of an additional party (the transit country) makes the question of how the rent is captured less straightforward than the method discussed above. The key actors are the company (or companies) involved in the project (which have to justify their costs and establish a profit margin for the project to remain viable), the government of the producing country (which seeks to develop its resources for both domestic energy requirements and export, and also to earn revenue), and the government of the transit country. Without the transit country, the pipeline project either will simply not run or, if there are other possible transit routes, will be more expensive because the costs of the alternatives are likely to be higher (project planning is logically expected to proceed first with the most economic – or cheapest – option).

In cross-border pipeline projects involving transit, the producer government pays a transit fee to the transit country, or the transit country receives an off-take of the commodity in agreed fractions, or a combination of both occurs. The general objective of the transit country can be said to be similar to that of the producing government, in terms of capturing rent from the project. The problem, however, lies in three key issues:

▸ how the transit country justifies seeking to capture this rent;
▸ how the rent is actually determined; and
▸ how the transit country's fee or off-take is worked out with the least potential for future dispute.

The involvement of the transit country in the pipeline project can be active (when it owns and maintains the stretch of pipeline that runs through its territory, which could imply the pre-existence of a domestic pipeline network) or less active (when its role does not go beyond allowing the pipeline to run across its territory and collecting off-take from the pipeline). In the former case, one could argue that a transit fee is clearly justifiable. The transit country runs and operates an existing pipeline, which connects the producer country to the destination market, and it incurs costs while so doing. Therefore, it should be allowed to recover its costs (which in this case are quite tangible), and should be allowed a profit element (which can add complexities in terms of the method of determining the country's costs of capital). The transit fee should at least reflect this. There is also an opportunity cost factor: if the transit pipeline is part of a domestic network, for example, the transit country is right to seek compensation for the 'distortions' to the domestic market. This can be worked out as the opportunity cost of the 'lost gains' in domestic trade. When the transit country's involvement is less active, there are two sides to the argument. On the one hand, it can be argued that collecting the off-take for its domestic energy requirements is fair compensation to the transit country; the country benefits from cheaper access to sources of energy from the producer country, and the opportunity cost of seeking alternative sources of energy can be reflected in the off-take proportions. The transit country is, therefore, not justified in seeking further rent from the project. On the other hand – in line with the views on the purpose of the transit fee discussed in section 2.5.2 – it can be argued that the transit fee should reflect one or a combination of such factors as loss of sovereignty, contribution to value added, and being the cheapest alternative. However, the entire pipeline project is valueless without the transit country, or more expensive with alternative routes, which suggests that the transit country can – and will, in some cases, because it can – seek to capture more rent from the project via transit fees or off-take. It is inevitable that the transit country will seek a fee as long as the project will produce rent, whatever the justification. Whether it is reasonable or not will then depend on how the rent is shared.

The other problem – how the transit country can seek to capture rent, and how it does so with the least potential for dispute (or no dispute at all) – is not straightforward. The producer government seeks to capture rent through such means as bonuses, royalties, and taxes. Assuming the rent is based on the difference between the value and costs along the entire value chain, the producer government will try to capture all of the rent available for the project. Therefore, it is not unreasonable to suggest that it is from this captured rent that the producer government should pay the transit fee to the transit country. The question of what proportion of this rent will be paid as a transit fee then depends on the relative bargaining power of the parties to the negotiation, which in turn raises even more questions, such as how it can be ensured that the party with greater bargaining power does not continually squeeze for more rent.

2.6 Conclusion

It has been established that so long as a project earns rent, the transit country will seek to capture as much as it can from this rent. Also, host governments capture rent from the project through royalties, bonuses, production-sharing, or taxes. The transit country captures this rent with the transit fee. The question, however, is how the transit fee is worked out; what proportion of this rent does the transit country get, such that the likelihood of the transit country disrupting the operating pipeline in its quest for a bigger share is reduced or removed? It has become acceptable in the scarce literature available on the subject to leave the determination of how this rent is shared to a case-by-case basis. However, a number of factors should generally apply in the determination of the basis for the transit country's fee. These factors include the following (in no specific order):

- the costs to the transit country;
- the value of the transit route;
- the availability of alternative transit routes; and
- the relative bargaining power of the parties involved (companies, producer government, transit country).

These factors are analysed in the following chapter. The eventual transit fee should ideally (for calculation purposes) be a function of volume and pipeline dimensions (not as defined by Masseron, 1990) – say, in US dollars per unit volume of oil or gas traded. The conclusion to draw from this chapter regarding the implications of the economics of pipelines and of cross-border and transit economics for the transit fee is one of certainty in certain respects, and ambiguity in other respects. The concept of economic rent clearly explains the rationale behind the transit fee. As long as the project earns rent, the rent-seeking tendency of the transit country is inevitable. What is unclear at this point, however, is how the rent should be shared (how the transit fee should be set) such that all parties are satisfied (in other words, whether any rent-sharing formula can indeed be reasonable, objective, transparent, and non-discriminatory in the long term). The major problem with the general determining factors of the transit fee listed above is attaching a value to them for estimation purposes. There is a possibility that the entire transit fee formation process will boil down to the bargaining process and to the bargaining abilities of the parties involved, especially of the transit country.

Notes

1. Masseron (1990) suggests that the transit tariff includes a fixed-cost component and variable costs proportional to the length of the pipeline.
2. A detailed discussion of these factors is not directly relevant to this research and is thus not included. They are, in any case, self-explanatory.

3 These factors would directly or indirectly influence a number of decisions, such as the pipeline route and project lifespan (based on reserve size calculations), and are also factors in the determination of the costs of the project.
4 Exploration and production (E and P) costs become unclear and uncertain only if the exploration venture is unsuccessful; 9 out of 10 such ventures are (Johnston, 1994: 6). However, this study assumes that E and P is successful and that the oil or gas is available for transportation to the market via pipeline.

3
The Role of Bargaining in Oil and Gas Transit Pipelines

Abstract: *Given the rent-capturing behaviour of the transit country, which is enhanced significantly by its improved position in terms of the obsolescing bargain, this chapter reviews critically the basic principles of bargaining theory. The objective is to explain the role of bargaining in the determination of transit agreements for cross-border oil and gas pipelines. More specifically, this chapter investigates the sources of bargaining power, how the balance of power shifts along the pipeline supply chain, and how (or whether) bargaining power differs with respect to the commodity traded (i.e. oil or gas).*

Omonbude, Ekpen James. *Cross-border Oil and Gas Pipelines and the Role of the Transit Country: Economics, Challenges, and Solutions.* Basingstoke: Palgrave Macmillan, 2013. DOI: 10.1057/9781137274526.

3.1 Introduction

As concluded in the previous chapter, the question of what is a reasonable transit fee chargeable by the transit country cannot be resolved objectively. The transit country can seek to optimise rent from the project so long as the project earns rent, subject to constraints identified in Chapters 4–6. The question of how this rent is apportioned among the players – especially how the transit country charges its transit fee – is of relevance to this research because it affects the answer to the question of whether transit fee agreements can indeed bear the characteristics of reasonableness, objectiveness, transparency, and non-discrimination.

Given the rent-capturing behaviour of the transit country, which is enhanced significantly by its improved position in terms of the obsolescing bargain, this chapter reviews critically the basic principles of bargaining theory. The objective is to explain the role of bargaining in the determination of transit agreements for cross-border oil and gas pipelines. More specifically, this chapter investigates the sources of bargaining power, how the balance of power shifts along the pipeline supply chain, and how (or whether) bargaining power differs with respect to the commodity traded (i.e. oil or gas).

This chapter relates the bargaining principles of risk aversion, commitment tactics, inside and outside options, patience, and information asymmetry to the way in which bargaining power shifts among the parties to a cross-border pipeline agreement involving transit through a third country. In addition, it explores the role of investment in pipeline capacity and its effect on the shift in bargaining power.

3.2 Bargaining theory – a literature survey

This section surveys the literature on bargaining theory. The objective here is to draw out the salient issues about the roles of the key players in the bargaining process and to point out factors that influence these roles (further discussed in section 3.3, which outlines a non-technical approach to bargaining for the purposes of this research).

There are basically three bargaining problems: the outcome of the negotiation, the process of concession, and how the basic parameters of the bargaining situation can be influenced or changed by the negotiators (Cross, 1969). The first problem (the outcome) includes such parameters as the preferences of the parties involved in the negotiation process, the benefits to be distributed, the maximum possible payoff for each party, and the payoff values associated with a state of disagreement between the parties (Pen, 1952). With regard to the process of concession (the second problem), Cross (1969) argues that the bargaining outcome can be specified or specifically predicted if the process is set out in a deterministic manner. Cross further argues that so long as the focus is on bargaining behaviour and the variables that contribute to such behaviour, no general predictions can be made about the outcome. This approach clarifies the importance of different tactics and shows how changes in the behaviour of the parties involved (if such changes are not too complicated) can affect the outcome of a bargaining situation. The third problem simply considers the influence of the parties to a bargaining situation with respect to the factors that determine the outcome. How the roles of the key parties to a pipeline (especially the producing country and the transit country) influence the bargaining process constitutes a major topic of this book.

The bargaining process is fundamentally time dependent. The passage of time has a cost in terms of currency and utility sacrificed by postponing consumption. This cost is argued to be what motivates the entire bargaining process.

Game-theoretic solutions to the bargaining problem, as championed by Nash (1953; Roth, 1978, 1979), focus on the first approach discussed in this section – that is, on the outcome rather than the process. The Nash solution corresponds to the solution in which the parties to the bargaining situation make equal proportional sacrifices. It can be expressed in terms of elasticities. The Nash solution is argued in some of the literature to satisfy a number of conditions such as efficiency (or Pareto optimality), individual rationality, and feasibility. The Nash solution to bargaining is argued to be symmetric and to be unaffected by changing the available agreements, so long as new agreements are not preferred to the original solution. These conditions are often referred to as Nash's *axioms of choice*.

It is argued that the major advantage of this solution to the bargaining problem is that it helps to simplify complex social situations. However, one criticism is that this solution is unable to narrow the range of possible solutions far enough to establish much explanatory or predictive power (Esteban and Sakovics, 2003).

Other approaches to the bargaining problem are more process-focused. It is argued that a state of conflict will yield a known payoff utility to each of the parties, and that these values are unchanged throughout the bargaining process. The implication is that these values are independent of time and demands made by the parties (or attempts by the parties to change the factors that could determine the bargaining outcome).

Certain general points can be picked out from the literature on bargaining for the purposes of this book. While some authors focus on the bargaining outcome, others pay attention to the actual bargaining process (which may or may not include attempts by the parties to the negotiation to influence the process), but the seminal works on the subject do not appear to have considered both process and outcome. Another important feature of these works (and, to some extent, a practical disadvantage in terms of simple interpretive applications) is the overly technical and complex nature of the subject. More recent research has attempted to simplify matters by identifying the relevant aspects of the bargaining situation. In so doing, there is a combined discourse of both process and outcome, discussed in detail in the following section.

3.3 The concept of bargaining in economics – a non-technical approach

An exchange situation in which individuals or organisations can engage in mutually beneficial trade but have conflicting interests over the terms of trade is defined as a bargaining situation. Generally speaking, a bargaining situation is one in which two or more players (individuals, organisations) have a common interest to cooperate but have conflicting interests over exactly how to cooperate (Osborne and Rubinstein, 1990). There are two major motives for interest in bargaining situations:

- Many important and interesting human (economic, social, and political) interactions involve bargaining. It is ubiquitous.
- An understanding of bargaining situations is fundamental to the development of the economic theory of markets (Muthoo, 1999).

The main problem that confronts the parties to a bargaining situation is the need to reach agreement over exactly how to cooperate. Each party/player would like to reach some agreement rather than to disagree and not reach any agreement (although it can be argued that it is possible for a party not to enter into agreement at all); however, each would also like to reach an agreement that is as favourable to it as possible. It is thus possible that the players will strike an agreement only after some costly delay, or indeed fail to reach any agreement.

The following sections conduct a non-technical discussion of six key principles of bargaining as defined by Muthoo (1999), and these principles are then related to transit agreements for cross-border oil and gas pipelines to explain the relevance of bargaining. The factors from which these principles are developed are as follows:

- a common interest to trade (the bargaining pre-requisite);
- patience;
- risk aversion;
- inside and outside options;
- commitment tactics; and
- information.

3.3.1 Bargaining theory and cross-border oil and gas pipelines

The objective of the following sections is to draw on the principles of bargaining to explain the bargaining role played by each key party to a cross-border transit oil and gas pipeline project. By way of recapitulation, there are four key parties to transit fee agreements (for the purposes of this book): the company operating the pipeline, the producer country, the transit country (both pure transit and off-taker), and the destination country (the market or connection point to other markets). It should, however, be noted that the final destination could be the high seas, in which case the role of destination countries is weakened at best, and more likely non-existent. This chapter, therefore, concentrates on the producer and transit countries.

The main focus of this research is the transit fee, and, therefore, the transit country. Therefore, treatment of the bargaining outcomes influenced by the factors discussed here is restricted to the transit fee and thus transit country behaviour. For the purpose of this analysis, the bargaining outcome in this regard is one of three:

- an acceptable fee or off-take to the transit country, with no likelihood of the transit country seeking contract renegotiations during the operation of the pipeline;
- an acceptable fee or off-take to the transit country, but with the likelihood of the transit country taking advantage of the obsolescing bargain; or
- an unacceptable fee or off-take to the transit country.

The role of bargaining theory and its implications for these outcomes are revisited in section 3.4.

3.3.2 A common interest to trade: the bargaining pre-requisite

According to Muthoo (1999), bargaining situations will not arise without a common interest to trade shared by the players. This is the fundamental basis for trade, and trade is the fundamental factor in any bargaining situation. Parties involved in a bargaining situation are better off if one party values his commodity at a certain price, the second party is willing to pay an amount higher than this valuation, and trade occurs at a price between the two parties' values. However, the two parties have conflicting interests over the price at which to trade.

The bargaining pre-requisite – in terms of applicability to transit pipelines – is straightforward. It can be argued that the benefits from oil and gas pipeline projects will generate interest on the part of the transit country to deal with the producer country. However, the transit country – as already established – will seek to maximise its share of the available surplus from the pipeline in fee or off-take terms; thus the question of how to reach an agreement persists.

3.3.3 Patience and bargaining

An important factor in this bargaining principle is the cost incurred from haggling. One basic source of this cost is the time it takes to haggle, and time is usually of some importance to each party involved. The key principle here is that a party's bargaining power is higher the more patient the party is in comparison with the other party. If there is no cost to the negotiating parties of continuous bargaining, then both parties will continue to seek agreement on their respective terms (Muthoo, 1999). This is, however, not typically the case in real bargaining situations, as time is usually a key factor for both parties.

Example

> There are two parties to a negotiation – A and B. A has a good (X) to sell; B wishes to buy the same good. The price for X will be higher if B is less patient than A. Therefore, A can be said to have the bargaining power. If, on the other hand, A is desperate to sell X, and B is more patient, bargaining power will shift to B.

It is argued in some of the literature that if it did not matter when the negotiators reached agreement, then it would not matter whether they reached agreement at all. This principle is commonly demonstrated in wage negotiations. Also, it is common that poorer parties are more eager to reach agreement in any negotiations (Lindblom, 1948). The poorer party is usually more desperate. In cases of long-term unemployment, for example, many unemployed people will be willing to accept work at almost any wage. This high level of impatience can be exploited by employers, who may thus obtain the gains from employment. Therefore, because it induces a higher degree of impatience, poverty adversely affects bargaining power. This is reflected in international trade negotiations, in which richer countries tend to benefit more than the poorer countries. However, it is also possible that the poorer party has more bargaining power, as it can be argued that this party has little to lose. A poor and unstable government, for example, could have less consideration for its international reputation, and as such might seek to capture as much rent as it can without concern for the negative implications of its action.

The time factor and bargaining positions

For cross-border transit oil and gas pipelines, it is always important that construction and maintenance schedules, pipeline capacities, and off-take portions are adhered to, especially by the exporter. Timing is important to the successful construction and operation of trans-boundary pipelines, and so also is the security of supply.

For the pure transit country, timing plays a significant role, in that it appears to give it the bargaining upper hand. The general argument about patience and bargaining is that the party for whom prolonging negotiations costs less will continue to do so, thereby exploiting the relative impatience of the other party. This implies that the pure transit country – on the face of it – does not necessarily stand to be worse off from disrupting the pipeline in its bid to increase its share of the rent through raised fees.

The position of the off-take transit country is slightly different from the pure transit case. Depending on the value of the oil or gas to its domestic economy, it could stand to lose out by interrupting supply in its attempt to extract more rent from the project. On this basis, in terms of patience and bargaining, it does not necessarily cost the off-take transit country less to prolong negotiations for higher fees or off-take quantities, so it does not enjoy as much bargaining strength as the pure transit country does in such a situation. However, the transit country could act because of a change in policy (see section 3.3.6 on commitment tactics) or a deterioration in diplomatic relations between itself and the producer country, the destination country, or both. In such a case, bargaining power shifts in favour of the transit country.

For the producer country, however, timing is of critical importance. The producer country is keen to ensure prompt and secure supply of the oil or gas to its markets. The cost of delays or interruptions to supply could be huge; therefore, the producer country is worse off in terms of bargaining power than the pure transit country. In negotiations with the off-take transit country, however, the producer country can exercise some of its bargaining power in the knowledge that the off-taker depends on the oil or gas from the system. It is useful to note that the off-take country could siphon off higher quantities than contractually determined (as is the case with many pipeline projects, with or without justification).

3.3.4 Risk aversion

There exists a risk of breakdown in negotiations in any bargaining situation. According to Muthoo (1999), negotiations could break down into disagreement as a result of exogenous factors such as:

- impatience or frustration at the protraction of negotiations (which is random behaviour); or
- intervention by a third party (the effect of which is the removal of the gains from cooperation that exists between the players).

The more risk-averse party is more eager to minimise the risks of breakdown in negotiations. Given that the less risk-averse party demands a larger share of the net surplus, he can, therefore, exploit the more risk-averse party's eagerness to trade. If both parties are equally risk-averse, *ceteris paribus*, then they are likely to split the surplus equally.

Example

> Continuing with the example used in section 3.3.3, good X could be a computer. Improvements in technology, such that there are newer, cheaper, more efficient versions of X, mean that B could go elsewhere to purchase it. B, therefore, has more bargaining power. The scenario can also be reversed. If X is property, for example, and property prices go up in a booming market, A can sell elsewhere, thus shifting bargaining power to A.

The key principle here is that a party's bargaining power is higher the higher is his payoff following the occurrence of the exogenous and uncontrollable event that triggers a breakdown in the negotiations. In other words, a party's bargaining is lower the higher is the other party's payoff in the same eventuality.

The exact distribution of the net surplus between the parties will depend on their relative degrees of impatience as well as their risk-aversion levels. Therefore, a party's share of the net surplus is smaller the more averse to risk he is than the other negotiator.

Risk-averse positions of the producer and transit countries

The more risk-averse party is willing to minimise the risk of a breakdown in negotiations, as shown above. For the transit country, a number of economic and political factors determine how the risk-aversion principle affects its bargaining position. These factors include the following:

- economic dependence on oil and gas revenue;
- political/diplomatic relations between the countries involved as well as relations with international institutions; and
- contribution to the cost of the project.

Economic dependence on oil and gas revenue

For the off-take transit country, the implications are straightforward. If the country's economy will benefit immensely from the transit fee (or if the transit fee is an important source of revenue to the country) and if its domestic energy market is significantly dependent on off-take volumes of oil or gas from the pipeline, then it is less prepared to risk losing this income. Therefore, it is less likely to exercise its bargaining power once production and transportation commence, implying that it is risk averse by virtue of its reliance on oil and gas off-take. This is, of course, subject to *ceteris paribus* conditions (e.g. no alternative sources and no alternative pipelines). However, in the case of the pure transit country which will not necessarily be worse off without the earnings from the pipeline project, not only can it afford the time to drag out negotiations (as discussed in 3.3.3), but it is also less risk averse than the other parties (especially the producer country).

As shown in the case studies in Chapter 4, the producer countries involved in the current major pipeline projects (especially in the Caspian region) depend heavily on revenue from their oil and gas reserves. The importance of the pipeline projects to these countries is, therefore, high: they are keen to exploit their resources. For this reason, they can be argued to be more risk averse than the transit country, such that the transit country has more bargaining power. The shift in bargaining power (pertaining to the apparent risk-averse position of the producer country) will depend on the type of transit country (or countries) involved (offtaker versus pure transit).

Political/diplomatic relations

The implications of good diplomatic relations among the countries involved (as well as relations or affiliations with international institutions) for the level of risk aversion of the producer and transit countries are straightforward. The usual outcomes of positive diplomatic, political, and economic relationships include:

- shared, compatible economic goals;
- mutual dependency (this topic is developed further in Chapter 6);
- compatibility of legal and regulatory regimes; and
- less likelihood of political and economic dispute, which implies less likelihood of disruptions to the pipeline.

If the key aspect of the risk-aversion principle in bargaining theory is the risk of a breakdown in negotiations, mutually beneficial diplomatic relations between the parties to a pipeline agreement will minimise this risk. From the above discussion of the role of economic dependence on the project in determining the risk aversion of the countries involved, the problem would appear to be with the pure transit country (in terms of the strength of the pure transit country's bargaining power). Mutually beneficial diplomatic relations between the countries could reduce this imbalance.

Contribution to the cost of the project

Parties involved in meeting the costs of a project obviously have a direct interest in the project. What this implies in relation to the levels of risk aversion of the parties (especially the transit country) is that there is less likelihood of breakdown: the more direct involvement (or responsibility) a party has in the cost aspects of the project, the more risk averse the party will be.

3.3.5 The role of inside and outside options

This principle relies on two assumptions: first, that there is no risk of a breakdown in negotiations and, second, that each player values time equally. A party's outside option will increase his bargaining power if and only if the outside option is sufficiently attractive (Muthoo, 1999; Koskela and Stenbacka, 2000). An unattractive outside option will have no effect on the bargaining outcome. The credibility of the outside option is, therefore, crucial to the bargaining outcome. The outside option can be a very effective bargaining tool if it is better than the negotiated offer for trade.

Example

In the negotiation for a price for X, A could receive an offer from another potential buyer, C. The shift in bargaining power will depend on:

- whether the price offered by C is higher or lower than B's offer;
- whether C's offer is the only outside option; and
- whether C's offer is credible.

If the price offered by C is higher than B's (as well as credible), then A has the advantage of exercising its outside option, implying greater bargaining power over B.

A party's inside option comes about from a payoff obtained during the bargaining process (i.e. when the parties are in temporary disagreement). The key principle is that a party's bargaining power is higher the more attractive is his inside option, and the less attractive is the other party's inside option. This principle holds only if both parties' outside options are sufficiently unattractive. If, on the other hand, one party's outside option is sufficiently attractive, the inside options of both parties will have no effect on the bargaining outcome. The negotiator with the more attractive outside option simply gets the more favourable deal. If both parties' outside options are sufficiently attractive, it is mutually beneficial for both parties to exercise them. The impact of this on the bargaining outcome would thus depend on such other factors as each party's risk-aversion level or patience.

Example

If X is a house, the seller (A) might still be living in it. A's inside option is the utility from living in the house. A's bargaining power will increase in proportion to this utility, because A will thus be less eager to sell the house. In this case, the principles of patience and risk aversion will come into play.

In terms of the applicability of this principle to transit pipelines, it is helpful to define the inside and outside options for the producer and transit countries.

Options for the transit country

In the case of the transit country, the outside options are clear. One of these outside options is no pipeline running through its territory at all. This is most applicable to the pure transit country. The transit country – on the basis of the argument that it will not be worse off without the pipeline – can exercise its outside option to enhance its bargaining power in negotiating the transit fee. The other outside option available to the transit country is for it to participate as a transit route for another pipeline project. Again this is a *ceteris paribus* condition. The alternative project must be commercially viable and have enough throughput to meet the transit country's off-take requirements. This also, of course, assumes that an alternative pipeline project intent on using the same country as its transit route does exist.

For the pure transit country, this option can be exercised as a straightforward bargaining tool. For the off-taker, however, exercising this outside option would depend on such factors as domestic energy requirements compared with the expected throughput and capacity of the two pipeline options, diplomatic relations with the country from which the other pipeline originates (or with the destination country), and the cost implications of choosing to participate as a transit route for one pipeline over the other. In other words, exercising this outside option depends significantly on its credibility. The inside option for the transit country is also straightforward, and as such does not require further discussion. It is the potential benefit that will accrue to it either in the form of the transit fee or as off-take volumes for domestic use (or both).

Options for the producer country

The obvious outside option available to the producer country is an alternative transit route. As discussed in Chapter 2, in relation to the implications of pipelines as natural monopolies, the ability of the producer country to exercise this outside option is heavily dependent on the economic viability of the alternative pipeline route. Since the entire planning of pipeline projects is usually subject to rigorous financial appraisal before transit routes are selected, it is likely that the alternative pipeline route is more expensive. The implication of this for the producer country's bargaining position is simple: the obsolescing bargain.

Another option for the producer country (however unacceptable in international law and politics) is force. As this option is not unrealistic, it warrants mention in this analysis. The producer country could exercise its military might over the transit country to secure the passage of the oil or gas via the pipeline. An example (not directly related to oil and gas transit pipelines) can be taken from the prospects of US oil imports during the first and second oil shocks.[1] In a 1975 article written under the pseudonym Miles Ignotus, details of how to use military force against Saudi Arabia were outlined, with the objective of breaking the Organization of Petroleum Exporting Countries' 'control' over oil prices during that period. Although the effects might not be approved of by the international community, the country with more military might certainly has a bargaining advantage. If the country is an exporter, then the bargaining position of the transit country is muted. This angle is, however, not pursued further in this book.

3.3.6 Commitment tactics

In some bargaining situations, there is a tendency for the parties involved to take actions before or during negotiations that partially commit them to their deliberately chosen bargaining positions. They are referred to as *partial* commitments because they can be revoked. There are, however, costs to revoking these partial commitments. Muthoo (1999) suggests, as a principle, that a party's bargaining power is enhanced by a higher cost of revoking its partial commitment; that is, the higher the cost of revoking its partial commitment, the more the party is committed to its chosen bargaining position. The high cost of revoking such commitments is argued to be a source of strength in bargaining situations. In international trade negotiations, for example, a government's bargaining power is higher because of its perceived high cost of reneging on public commitments to its electorate. To explain this principle, Muthoo uses an example from Schelling (1960: 27), which can be summarised as follows.

Example

> A typical example of the employment of the commitment tactic is in wage negotiations between trade unions and employers. It is not uncommon for union officials to convey the impression that the consequence of backing down from their demands is an inevitable

strike by the workforce (i.e. to convey the impression that they no longer have control of their union members). This could force the hand of the management, implying that bargaining power could then shift to the labour union.

The deployment of this tactic in negotiations could be complex, considering its strong potential to lead to disagreements or costly delayed agreements, especially when the stance of the negotiating party is strongly influenced by high costs of revoking such commitments. In situations where these partial commitments (or demands) are irreconcilable and the costs of revoking them are significant for both parties, the risk of a breakdown in negotiations is very high. If the cost of revoking these commitments is less for one party than for the other, then it is more likely that the party with the lower cost will back down to strike a negotiated settlement. In negotiations between a democratically elected government and a dictatorship, for example, it is argued that the negotiators from the democratic government would have more bargaining power, as there would be little or no cost to the dictatorship of revoking commitments, while the other government could be a minority in a very strong democracy.

Commitment tactics in the context of transit pipelines

The parties to a transit pipeline project are (ideally) committed to the relevant agreements governing construction, operation, maintenance, and rent-sharing. The producer country, for example, is committed to selling oil or gas to the market (recipient country) for relatively long periods (roughly between 15 and 30 years, as shown in the case study chapter). The producer country is also committed to paying a fee to the transit country or agreeing certain off-take volumes of oil or gas for the transit country. In the case of the transit country, its expected obligations to transit pipeline agreements include allowing right of passage for the pipeline itself, a commitment not to exceed agreed off-take volumes (unless stated in the contract), and meeting any cost obligations to the project. The key questions regarding the applicability of the commitment tactic principle to transit pipelines are whether these commitments can be argued to be partial (whether they can, therefore, be revoked) and what it would cost the parties to revoke these commitments.

The producer country – it would seem – would struggle to revoke its commitment to the buyer after agreeing to a contract. There would be serious revenue implications for the producer country's exploration and sale of the oil or gas, depending on its reliance upon revenues from the sale of the hydrocarbon in international trade, and thus there is an implicit commitment to its domestic economy. There are, of course, situations that could necessitate the termination of such agreements, but doing so would not be without serious cost implications for the producer country. For the pure transit country, it would appear as though there is not much – by way of cost of revoking its commitment – to lose (again, reference is made to the argument that the pure transit country will not necessarily be worse off without the pipeline passing through its territory). If, however, the pure transit country attaches value to the compensation received via transit fees (which might be reflected in its fiscal policy, for example), a partial commitment is created, and the costs of revoking the commitment change. Regarding the off-taker, whether it can revoke its commitment will depend on factors discussed in the previous sections on risk aversion and inside/outside options. A clear dependence on the off-take volumes from the pipeline automatically establishes a commitment by the transit country to its domestic consumers. An example is when the transit country continues to off-take without paying, or when it siphons more from the pipeline than contractually determined.

On the basis of the theoretical underpinnings of this bargaining principle, and limiting the commitment context simply to the pipeline agreements, there is a problem with the theory: it will not apply exactly to transit pipelines. If the theory is that the party with the higher cost of revoking its partial commitment is likely to possess more bargaining power, the producer country in this context should have more bargaining power than the transit country (both pure transit and off-taker). This does not, however, seem to be the case, since the transit country can disrupt the transportation of oil or gas through the pipeline for the simple reason that it costs less to revoke its commitment to the project. The theory would appear to work inversely in this case. If the transit country is more committed to the project (i.e. the transit country is tied to other diplomatic or economic commitments), then its costs of revocation are higher, and the implication of this is a weakened position in terms of bargaining power (an idea which is further developed in Chapter 6).

3.3.7 Information and bargaining

The extent to which information about various factors is known to all the parties in a bargaining situation is an important determinant of the bargaining outcome. It is argued that one of the consequences of information asymmetry in the bargaining process is that an agreement may not be struck, even when it is favourable for both parties to agree (Muthoo, 1999; Sen, 2000). The absence of complete information in the bargaining process usually leads to inefficient bargaining outcomes – indicated by disagreements or costly delays – but the better-informed party enjoys more bargaining power. Such costly delays can be used as a mechanism to enhance bargaining power; previously unavailable information (or privately held information) eventually becomes available to the uninformed party.

Information asymmetry and transit pipelines

For transit pipelines there are two sides to the application of this principle, depending on the availability of an economically viable alternative pipeline or pipeline route. If there is no alternative, then the question of enhanced bargaining power as a result of information being unavailable to other parties does not arise, as modern oil and gas pipeline projects do not necessarily benefit hugely from withholding information – the costs, fees, and revenue streams are known (if not to the general public) to all the relevant parties. On the other hand, if there is an alternative, and given that such issues as transit fee determination are not transparent in most cases, information asymmetry could apply.

3.4 The bargaining outcome

Section 3.3.1 briefly outlined the three possible bargaining outcomes for the transit country and the transit fee with regard to oil and gas transit pipelines:

- an acceptable fee or off-take to the transit country, with no likelihood of the transit country seeking contract renegotiations during the operation of the pipeline;
- an acceptable fee or off-take to the transit country, but with the likelihood of the transit country taking advantage of the obsolescing bargain; or
- an unacceptable fee or off-take to the transit country.

What the discussion in section 3.3.1 did not do was distinguish between two different bargaining situations regarding transit agreements for oil and gas pipelines. The first – the *pre-pipeline* bargaining situation – would lead to an outcome of agreement on the terms of the contract before the actual construction and operation of the pipeline commenced. The second – the *operating pipeline* bargaining situation – is different from the first in that the implications of interruptions are more severe than in the pre-pipeline situation, taking into consideration the investment in the construction and operation of the pipeline. The first and third possible outcomes can, therefore, be said to be pre-pipeline bargaining outcomes, and the second an operating pipeline bargaining outcome. The key question here is how the first outcome can be maintained, or how the second outcome can be prevented.

Section 3.3 defined bargaining principles in the context of transit pipelines. The objective of this section is to adapt these principles to determine the possible pre-pipeline or operating pipeline bargaining outcomes. Seven influencing factors have been identified for this purpose (Figure 3.1):

▸ the common trade interest;
▸ off-taking by the transit country;
▸ the economic contribution of the transit fee;
▸ the availability of alternative pipelines or pipeline routes;
▸ political and diplomatic relations;
▸ the relevance of the project to the domestic market; and
▸ active participation in cost-sharing.

These factors are inter-linked, and are discussed in the following sections.

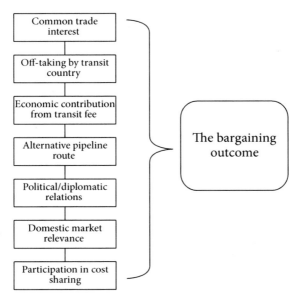

FIGURE 3.1 *Factors influencing the bargaining outcome*

3.4.1 The common trade interest

A common trading interest is particularly conducive to a positive bargaining outcome. Maintaining this interest to trade is especially important in the operating pipeline situation.

Mutual dependency between the countries involved in the transit pipeline project (the producer and transit countries especially) helps to ensure that the transit country does not seek to renegotiate once transportation commences. The common interest to trade must, however, stretch beyond interest in rent alone from the project. This can be argued to make the difference between a bargaining outcome in which the fee is accepted without likelihood of future renegotiation demands and one in which more rent will be sought by the transit country. It is, therefore, helpful if there are economic linkages between the parties involved, apart from the transit fee or off-take. This line of reasoning is developed further in Chapter 6 on mutual dependencies.

The benefit of maintaining this common trade interest between the parties to the pipeline project is obvious: less opportunity for disputes to arise, ensuring a secure supply of oil or gas through the pipeline. Such factors as economic and political stability play a key role in maintaining a common interest. If the economic and political relationship between the producer and transit countries is positive, it is more likely, *ceteris paribus*, that the operating pipeline bargaining outcome will favour security of supply for the exporter and importer countries. This does not rule out the obsolescing bargain, but the transit country will seriously consider the consequences of interrupting the pipeline before it seeks more rent from the project.

3.4.2 Off-taking by the transit country

The outcome in this case can also be argued to be positive, as the common interest to trade factor (above) shows, especially in cases where the transit country's domestic energy requirements are met by the off-take. The off-taking transit country is, *ceteris paribus*, less likely to interrupt the pipeline by renegotiating the terms of transit so long as it is reliant on the off-take volumes for its domestic energy requirements.

However, the possibility of the outcome changing in the future (once production and transportation commence) will arise if the transit country siphons off more oil or gas than contractually determined or fails to pay for the oil or gas it receives. Depending on what proportion of the volume of throughput beyond contractually agreed volumes is taken by the transit country, this would not necessarily immediately change the outcome, as the pipeline would continue to operate. However, the transit country will have to seriously consider the consequences of interrupting the pipeline. In a situation in which there are competitive consequences (i.e. if there is already an export pipeline or project in place), the transit country could seek more rent, stop paying for off-take, or siphon more volume from the pipeline (or a combination of all three). This can be done by the transit country to protect an existing export pipeline.

3.4.3 The contribution to the economy from the transit fee

The ideal argument pertaining to the effect on the bargaining outcome of the contribution to the transit country's economy of the transit fee would be that poorer countries dependent on foreign investment should be less likely to interrupt transportation, especially if the contribution from the fee is significant and there is little or nothing by way of alternative sources of foreign investment. For a poor pure transit country, the fee is a useful source of foreign investment. The transit country may not be worse off without this revenue source, but will certainly benefit from it.

There is the possibility, however, of a poorer country 'playing games'. The very fact that it is a pure transit country and will not be worse off without the transit fees suggests that there is room for it continuously to seek to squeeze rent from the project.

3.4.4 The availability of alternative pipelines or pipeline routes

If there are economically viable alternative pipeline routes or alternative pipelines of similar economic viability to the pipeline, the bargaining position of the transit country is limited. It is arguable whether the pre-pipeline bargaining situation will even arise, given that there will be no need for another pipeline if some alternative is equally viable. However, some projects are more politically driven than economically determined; therefore, when the pipeline is operational, such alternatives would affect the transit country's bargaining position.

The effect on the bargaining outcome is, in this situation, positive: the transit country is unlikely to renegotiate for higher fees or more off-take. The problem with this factor, however, is that since a pure transit country would not be worse off without the pipeline, it has nothing to lose from seeking renegotiations later in the life of the project. Moreover, off-takers in this position could be encouraged to cheat (i.e. to siphon more oil or gas than contractually determined).

3.4.5 Political and diplomatic relations

The argument here is that committing the transit country to relevant international institutions or objectives has a positive impact on the bargaining outcome; that is, it reduces the likelihood of further renegotiations. This is shown in the Baku–Tbilisi–Ceyhan and Shah Deniz pipeline case studies, where Turkey plays an important role in political and diplomatic relations. The objectives of the US are to unlock the Caspian reserves, avoid Russian or Iranian transit routes, and reduce traffic through the Bosphorus. One of Turkey's key objectives with regard to international trade relations is accession to the EU, so disrupting either pipeline through renegotiations would not enhance its international relations.

3.4.6 The relevance of the project to the domestic market

The relevance of the project to the domestic market can be linked to the factors discussed in sections 4.2 and 4.3. It is particularly applicable to the off-taker. In the absence of economically viable alternatives, and given a significant demand for the off-take for domestic energy use, the bargaining outcome will most likely be positive (i.e. agreement with no likelihood of future renegotiations). The bargaining position of transit countries in this scenario is one of risk aversion and commitment to domestic energy requirements.

3.4.7 Active participation in cost-sharing

Involving the transit country in sharing costs and, possibly, ownership or sponsorship of the project can also influence the bargaining outcome. This limits the bargaining power of the transit country because it is committed (or directly linked) to the value chain and, therefore, bears a share of the costs of any disruption to the pipeline (in proportion to its contribution to the project).

3.5 Bargaining outcomes for oil and gas pipelines – similarities and differences

Chapter 2 discussed a number of differences between oil and gas in the context of the pipelines used to transport them. The question in this section is whether such differences affect the bargaining outcome.

TABLE 3.1 *Similarities and differences in bargaining outcomes for oil and gas*

Principle/factor	Outcome
Common trade interest	No difference in outcome.
Patience/risk aversion	Effects of disruption are less for oil than for gas. Longer time and higher cost required to get gas transportation back on stream. Therefore risk of breakdown is higher for gas than for oil.
Off-taking/economic contribution	Greater rent to be shared from oil than from gas, due to huge cost differences. Implication could be more of the rent from oil being sought by the transit country.
Alternatives	Fewer alternatives in the case of gas due to huge costs of projects. Implication for outcome is more bargaining power to the off-taker from gas pipelines.
Politics/diplomacy	No difference in outcome.

Table 3.1 suggests that the bargaining positions of the countries involved in the project are not seriously affected or threatened by the type of hydrocarbon transported via the pipeline. Consideration of the key differences between oil and gas pipelines – such as the longer time it takes for disrupted gas pipelines to get back on stream, the fact that rent-sharing is driven more by contracts in the case of gas and by market mechanisms in the case of oil, and the fact that there is more rent from oil than from gas due to cost differences – leads to the conclusion that the bargaining positions of the producer and transit countries could be altered, but not significantly.

3.6 Conclusion

The primary objective of this chapter was to relate the principles of bargaining to the way in which bargaining power shifts among the parties to a transit pipeline agreement. The conclusion from Chapter 2 was that the transit country would continue to seek to capture rent from the pipeline simply because it could do so. Using bargaining theory to complement the analytical framework set out for this research in order to explore the determinants of the bargaining outcome with respect to oil and gas pipelines involving transit, this chapter has found that although the transit country would always seek to optimise rent, there are factors that could check its power to renegotiate. Some of these factors are market driven; others are regulatory or political.

It is important, during the pre-pipeline phase, that the bargaining outcome be mutually beneficial to all the parties involved. Within realistic constraints, transit agreements should be acceptable to the transit country in particular, to avoid cause for dispute when the pipeline is operating.

Linking bargaining principles to the possible outcomes of transit pipeline agreements (as done in this chapter) points in one particular direction: the significance of common interests among the parties involved, apart from the rent from the pipeline. This necessitates a further discussion (in Chapter 6) of how mutual dependency influences the nature of transit pipeline agreements.

Most importantly, the findings from this chapter and Chapter 2 are that there is no possibility of a cross-border oil or gas pipeline agreement that bears the characteristics of reasonableness, objectiveness, and non-discrimination, simply because the transit country *can* seek to disrupt the pipeline during operation for a bigger share of the rent. The bargaining positions of the parties involved are clearly not fixed (parties can influence the factors that affect their bargaining positions), and what is reasonable to one party is certainly open to debate by the other. If a 'reasonable' transit fee is defined within the context of one party having more bargaining power (however unfair this definition appears to the party with less bargaining power), then one may suppose that the outcome is reasonable at that time, until the determining factors shift bargaining power to the other party. This in itself implies that the transit fee cannot be reasonable, objective, and non-discriminatory.

Note

[1] For useful reading on the history of the oil market, oil shocks, and the economics of prices during this period see Yergin (1991: chs 29–34), Stevens (2000a), and Adelman (2002).

4
Bargaining Positions of the Parties to a Transit Pipeline: Four Case Studies

Abstract: *This chapter analyses the factors that influenced the transit fee agreements for the four main pipeline projects examined in this research. Specifically (and more importantly) it looks at factors that affected the bargaining positions of the parties involved in the pipeline agreements. The projects discussed in this chapter are the Baku–Tbilisi–Ceyhan oil pipeline, the Chad–Cameroon oil pipeline, the West African gas pipeline project, and the Shah Deniz gas pipeline project.*

Omonbude, Ekpen James. *Cross-border Oil and Gas Pipelines and the Role of the Transit Country: Economics, Challenges, and Solutions.* Basingstoke: Palgrave Macmillan, 2013. DOI: 10.1057/9781137274526.

4.1 Introduction

This chapter analyses the factors that influenced the transit fee agreements for the four main pipeline projects examined in this research. Specifically (and more importantly) it looks at factors that affected the bargaining positions of the parties involved in the pipeline agreements. The projects discussed in this chapter are the Baku–Tbilisi–Ceyhan (BTC) oil pipeline, the Chad–Cameroon oil pipeline, the West African gas pipeline (WAGP) project, and the Shah Deniz gas pipeline project.

The analysis is guided by the framework developed in Chapters 2 and 3, in which the economics of transit pipelines and the role of bargaining were reviewed and analysed. The question remains as to how the consequences of shifts in bargaining power among the parties to the pipeline agreement (i.e. the effects of disruptions to the pipeline) can be muted. Since the conclusion from the previous two chapters fundamentally suggests naked bargaining as the main solution to the potential problem of arbitrary transit agreement renegotiations, what this chapter does is search the case studies for characteristics that affect the bargaining positions of the transit countries. This chapter also points out factors that are useful in counter-balancing this bargaining power.

The background of the pipeline cases under review is established in this chapter, and relevant elements of the projects' history and development are identified. The implications for the pipelines of the economics and bargaining principles set out in Chapters 2 and 3 are then discussed. Findings from this chapter suggest a solution that identifies factors which are endogenous to the pipeline projects or agreements, but which require outside (or exogenous) factors to ensure success in the bargaining outcome.

In Chapter 1 it was briefly mentioned that four pipeline cases were selected for this research. Their selection was based on one simple but crucial criterion: publicly available and publishable information. The nature of cross-border oil and gas pipelines in the past has been such that there was little or no significant usable information on the basis of which to conduct detailed analyses of this nature. Table 4.1 summarises the transit pipeline cases analysed in the ESMAP study referred to in Chapter 1.

Table 4.1 analyses transit pipelines in terms of their response to three key factors that can be used to explain their success or failure. Success or failure is defined in terms of factors that could lead (or, in some of these cases, have led) to dispute. The success cases show that private sector involvement played a significant role, suggesting that a commercially

TABLE 4.1 Factors influencing the success or failure of cross-border pipelines

Pipeline	Response to factors		
	State interest	Adaptability	Rent-sharing mechanism
Success cases			
Transmed	Very high, common interest	Flexibility	Unclear, but apparently effective
Former Soviet Union/ Transneft	State ownership	Flexibility	Commercial constraints apparently effective
SuMed	No private sector commercial behaviour	Flexibility	Effective
Failure cases			
Iraqi Pipelines	No private sector commercial behaviour	No mechanism	Ineffective
Tapline	Private initiative, but in government context	No mechanism but eventual change	Ineffective
Recent pipelines			
Baku Early	Private initiative, but in government context	Dispute resolution mechanism exists	Balance of risk and reward
Maghreb–Europe	Private initiative, but in government context	Mechanism exists	Effective mechanism
Caspian Pipeline Consortium	State involvement	Mechanism exists	Effective mechanism
Canada–US	Entirely private	Maximum flexibility	Effective mechanism Competitive market
Bolivia–Brazil	Private, but state carried risks	Flexibility	Seemingly effective
Baltic pipeline	State involvement	Some flexibility	Potential for increased revenue – effective

Source: Adapted from ESMAP case studies. See ESMAP (2003) for further details.

driven project (in terms of ownership and/or sponsorship and available markets) is likely to cause fewer disputes than a purely state-driven project. In addition, the success cases reveal a degree of flexibility or adaptability to changes in the surrounding circumstances, ranging from changes in price to changes in government objectives. The value of the throughput (gas) was priced using inflation and indexation to oil prices. So, if oil prices went up or down by a certain proportion, gas prices – as far as the Transmed pipeline was concerned – reacted the same way. In addition, ownership of the Algerian gas passed to Italy once the gas entered Tunisian territory, so that any form of aggressive behaviour by Tunisia over transit terms would have been dealt with by Italy, not by Algeria. There is not much evidence from these cases to support the rent-sharing mechanism, but it is not unreasonable to deduce the effectiveness of such mechanisms in the success cases, and the significance of their absence for the long-term failures.

The failure cases exhibited the opposite characteristics in relation to these three factors. The analysis shows that the absence of a commercial motive, a mechanism that allows adaptation to changes in circumstances, and a transparent rent-sharing scheme will lead to disruptions to the pipeline.

In light of the analyses conducted in Chapters 2 and 3, it would be meaningful to apply the bargaining principles framework to these ESMAP cases. However, this is not possible because of the limited amount of publicly available information on these projects. The four projects studied in this book do not pose this challenge. The significant involvement at some stage of the World Bank and the International Finance Corporation (IFC) in these pipelines led to a considerable amount of transparency and publicly available data to analyse. What follows is a rigorous analysis of the four pipeline cases selected for this study in the context of the research problem and the analytical background so far developed in this book.

4.2 Basic details of the four pipeline projects

4.2.1 The BTC oil pipeline project

The BTC oil pipeline project is one of the six elements of the Azerbaijan–Georgia–Turkey pipelines system. The other elements are the Baku–Novorossiysk oil pipeline (also known as the Northern Route

Export Pipeline, NREP), the Baku–Supsa oil pipeline (or the Western Route Export Pipeline, WREP), the Baku–Tbilisi–Ezurum gas pipeline (or the South Caucasus Pipeline, SCP), the Azeri–Chirag–Guneshli (ACG) oil fields, and the Shah Deniz gas project. The BTC pipeline runs from Azerbaijan through Georgia to the Ceyhan terminal in Turkey, for further export by tanker to international markets. One of the main motives behind the construction of this pipeline was the inadequate capacity of the NREP and the WREP to meet demand from additional phases of the ACG oil fields. Another aim of the pipeline was to enable oil exports from the Caspian Sea without having to transit the Turkish Straits (argued to be environmentally sensitive and congested; IFC, 2003).

The first oil reached the Ceyhan terminal in May 2006, and the first tanker export of crude which had passed through the pipeline was in June of the same year.

The project is managed by the BTC Company sponsor group, which is made up of 11 national and international petroleum companies. BP is the major shareholder and operator of the pipeline, with a 30.1% stake. The other shareholders are the State Oil Company of the Azerbaijan Republic (SOCAR, 25%), Unocal (8.9%), Statoil (8.71%), Turkiye Petrolleri Anonim Ortakligi (6.53%), Total (5%), Eni (5%), Itochu (3.4%), ConocoPhillips (2.5%), Inpex (2.5%), and Amerada Hess (2.36%) – as shown in Table 4.2. The project is legally implemented in accordance with the terms of the inter-governmental agreement between the three countries (the Republic

TABLE 4.2 *Equity participation in the BTC Company*

Equity participant	Country	Equity (%)
BP (operator)	UK	30.1
SOCAR	Azerbaijan	25
Unocal	US	8.9
Statoil	Norway	8.71
Turkiye Petrolleri Anonim Ortakligi	Turkey	6.53
TotalFinaElf	France	5
Eni	Italy	5
Itochu	Japan	3.4
ConocoPhillips	US	2.5
Inpex	Japan	2.5
Amerada Hess	Saudi Arabia	2.36

Source: Adapted from www.bp.com.

of Azerbaijan, Georgia, and the Republic of Turkey) signed in 2000 and by the host government agreements.

The BTC pipeline was designed to have enough capacity to export all ACG volumes as well as production from other Caspian developments in the future. Also referred to as the Main Export Route Pipeline (or Main Export Pipeline), the pipeline is approximately 1,760 kilometres in length (442 kilometres in Azerbaijan, 248 kilometres in Georgia, and 1,070 kilometres in Turkey). It is expected to transport up to 1 million barrels per day (50 metric tons per annum). The pipeline starts at the Sangachal terminal near Baku and runs parallel to the WREP. At the Georgian point of connection, the pipeline diameter changes from 42 inches to 46 inches, and the pipeline runs westward to Turkey. At the Turkish point, the pipeline diameter reverts to 42 inches, and the pipeline runs south to Ceyhan ports. Throughout its length, the pipeline is buried at a depth of at least 1 metre, except for above-ground installations (the Sangachal and Ceyhan terminals, pumping stations, and so on).

4.2.2 The Shah Deniz gas field/SCP project

The Shah Deniz gas and condensate field is situated under the bed of the Caspian Sea, approximately 100 kilometres southeast of Baku, in water depths of up to 600 metres. Discovered in 1999, it is estimated to contain more than 400 billion cubic metres of gas reserves. The project's development plans include the installation of a fixed production platform linked by sub-sea pipelines to an onshore terminal.

According to the production sharing agreement (PSA) signed and ratified in 1996, the project is owned and managed by SOCAR, Turkiye Petrolleri Anonim Ortakligi, and four other national and international companies (BP, Elf, LukAgip – a joint venture of Lukoil and Agip, and Statoil; see Table 4.3). The project is legally implemented in accordance with the provisions of the PSA, the 2001 inter-governmental agreement between Azerbaijan and Turkey, an agreement between SOCAR and Boru Hatları İle Petrol Taşıma Anonim Şirketi, an inter-governmental agreement between Azerbaijan and Georgia, the host government agreements between the sponsor companies and the governments of Azerbaijan and Turkey, and the owners agreement. The Turkish section of the pipeline is owned by the Turkish government rather than the SCP partners.

Gas from this field is exported via an underground pipeline – the SCP – which runs parallel to the BTC pipeline from Azerbaijan and

TABLE 4.3 Equity participation in the South Caucasus Pipeline

Participant	Country	Equity (%)
BP	UK	25.5
Statoil	Norway	25.5
SOCAR	Azerbaijan	10
LukAgip (Lukoil and Agip joint venture)	Russia/Italy	10
TotalFinaElf	France	10
Oil Industries Engineering and Construction	Iran	10
Turkiye Petrolleri Anonim Ortakligi	Turkey	9

Source: Adapted from www.bp.com.

through Georgia to Turkey, where it is linked to the Turkish gas distribution network. The pipeline is 691 kilometres long, with 441 kilometres in Azerbaijan and 250 kilometres in Georgia, and is capable of carrying up to 7 billion cubic metres of gas every year (Crandall, 2006; BP website). According to BP, gas deliveries to Turkey commenced in September 2006.

4.2.3 The WAGP project

The WAGP covers about 1,033 kilometres, both onshore and offshore, from the Niger Delta region of Nigeria (Escravos) to its final destination terminal in Effasu, Ghana. The first portion of the pipeline delivers gas from Escravos to the industrial areas of Lagos, from the Alagbado distribution terminal. The Escravos–Lagos pipeline was commissioned in 1989 to supply natural gas to the Egbin power plant in Nigeria, in addition to other industrial consumers in the western part of Nigeria (Energy Information Administration (EIA) website).

The Escravos–Lagos pipeline already has capacity of about 900 million cubic feet per day of natural gas. From Alagbado, the WAGP runs 57 kilometres onshore to Seme beach in Lagos State, whence it continues offshore through Cotonou (Benin Republic), Lome (Togo), Tema (Ghana), and Takoradi (Ghana) to its final destination at Effasu. The pipeline's initial capacity will be 200 million cubic feet per day, with the capability to expand to 600 million cubic feet per day, pending growth in demand (EIA Country Analysis Brief).

The project was planned to initially transport 120 million cubic feet per day of gas to Ghana, Benin, and Togo in 2005. However, the construction of the offshore part of the pipeline between Nigeria and Ghana was completed only towards the end of 2006. It was originally estimated that gas deliveries would be approximately 150 million cubic feet per day on commencement of operation, reaching 210 million cubic feet per day after 7 years, and then 400 million cubic feet per day at full capacity (approximately 15 years after construction). In terms of demand, it is estimated that about US$600 million will be invested in the development of power facilities in the four countries for gas utilisation purposes. The first recorded delivery of gas via the pipeline was in December 2008, to the Volta River Authority in Ghana.

In terms of the legal structure, the governments of the four nations signed a heads of agreement in September 1995. The agreement summarised the principles that would guide the development of the pipeline. The legal framework for the WAGP project was established under the memorandum of understanding signed by the four countries. Other legal aspects of the project include the joint venture agreement naming ChevronTexaco as the WAGP project manager, an inter-governmental agreement, and administrative support from the Economic Community of West African States (ECOWAS). The agreement was signed by the four countries on the implementation of the WAGP. A treaty signed in February 2003 provides – over a 20-year period – the legal, fiscal, and regulatory framework, as well as a single authority for the implementation of the project.

The participants in the project include ChevronTexaco as the major shareholder and chief operator of the project, the Nigerian National Petroleum Corporation, Shell, the Volta River Authority, the Société Beninoise de Gaz, and the Société Togolaise de Gaz (Table 4.4).

TABLE 4.4 *Equity participation in the West African Gas Pipeline Company*

Participant	Country	Equity (%)
ChevronTexaco	US	36.7
Nigerian National Petroleum Corporation	Nigeria	25
Shell Petroleum Development Company	Nigeria	18
Volta River Authority	Ghana	16.3
Société Beninoise de Gaz	Benin	2
Société Togolaise de Gaz	Togo	2

Source: Adapted from www.wagpco.com.

4.2.4 The Chad–Cameroon oil pipeline project

The Chad–Cameroon Petroleum Development and Pipeline Project aims to develop the oil fields at Doba in southern Chad and then construct a 1,070-kilometre 30-inch pipeline through Cameroon to offshore loading facilities near Kribi on Cameroon's Atlantic coast. In addition to the development of the Doba oil fields and construction of the underground pipeline, the technical aspects of the project include related pumping stations, ancillary facilities, and infrastructure, and the installation of offshore floating storage, in addition to an offloading vessel at sea, associated marine pipelines, and other related facilities. The project sponsors are ExxonMobil (acting as the operator, with a 40% stake), Petronas (a Malaysian company, 35%), and ChevronTexaco (25%) (*source*: World Bank).

The considerable backing this project received from the World Bank was withdrawn in 2008, over Chad's failure to comply with elements of an agreement designed to channel revenue from the pipeline directly to poverty reduction. However, prior to the World Bank's withdrawal, the availability of a financial package from the outset of the project can be argued to have accounted for its relatively quick progress, as oil was first pumped through the pipeline in October 2003, almost a year ahead of schedule.

4.3 Key elements of the projects' history and development

This section analyses the key economic and political factors that came into play during the early stages of the pipeline cases studied here, with respect to the impact these factors had (or might have had) on the way transit fee agreements were (or could have been) made. For definitional purposes, the 'early stages' start with the pipeline project's conceptualisation.

Six factors have been identified for discussion, based on the pipeline economics and bargaining principles analysed in Chapters 2 and 3:

1. the objective behind the development of the pipeline project;
2. the cost of the project;
3. the economic and geo-political relevance of the project to the transit country;
4. the value of the transit route to the entire project;

5   the economic and political environment; and
6   factors relevant to the bargaining process.

4.3.1   The objective behind the development of the project

Pipeline projects can be classified as either commercially driven or politically motivated. As the case studies show, and as discussed in the following sections, in the case of commercially driven projects there is usually a ready market for the commodity (in the case of oil, should contractual problems ensue, the international market is competitive enough for producers to find buyers, *ceteris paribus*) or a long-term sales/purchase agreement (in the case of gas projects), and the costs of the project can be justified to the financiers on the basis of anticipated returns.

Politics alone cannot justify projects. Politicians regularly push bad investments. Of more interest is when political considerations affect decisions. In the less problematic cases, an ordinary cost analysis may ignore the cost of political risks. A more difficult situation occurs when the decision to choose an alternative route is based on a desire to punish a potential transit country.

With politically driven projects, a number of situations could occur. For example, the market could be either too small or non-existent, in which case the project would not be economically viable, as it could incur huge losses or barely break even. In such cases other factors, such as the development that will be generated from the project, are hastily put forward as justifications. Projects tend to comprise a mixture of commercial and political motives; there is a ready market, but diplomatic relationships between countries could be such that the choice of a longer and more expensive route for the pipeline is preferable. The Nord Stream pipeline from Vyborg in the Leningrad region of Russia to the coast of Germany is an example. While it has been argued to be economically viable, it was launched by the Russian Prime Minister at a time when an alternative route to Ukraine for future gas volumes was of political significance. In some cases there have been allegations that the motivation behind the project was more political than commercial. An understanding of such motives can be argued to influence the way the transit fee or transit arrangement is arrived at, as will be demonstrated in this chapter.

The BTC oil pipeline project

After the collapse of the Soviet Union in the early 1990s, it became apparent that there were vast untapped petroleum deposits in the Caspian region.

From a security of supply perspective, this region represented a potentially important alternative source of petroleum for Western markets. Building an oil pipeline from the Republic of Azerbaijan to Turkey was argued to further many US and regional policy goals, which included unlocking the Caspian reserves to the international market and ensuring that neither Russia nor Iran had a monopoly over pipelines from the Caspian region (Morningstar, 2003). The US was publicly opposed to any pipeline crossing Iran, as made clear in the Iran–Libya Sanctions Act at the time. Pipeline projects were thus started in the Caspian region, and the first two covered the routes from Baku to the Georgian port of Supsa and from the Tengiz oil field in Kazakhstan to the Russian Black Sea port of Novorossiysk.

The building of more pipelines between the Caspian Sea and the Black Sea was discouraged by both the US and Turkish governments, which raised environmental concerns: they feared that the tanker traffic generated by such projects would affect the already congested Turkish Straits.

The BTC Main Export Pipeline project thus arguably provided secure alternative access to Caspian oil for the Western market. With respect to a Russian monopoly over the route for Caspian oil, the BTC pipeline represented a competitive alternative to the NREP and WREP projects passing through the Turkish straits – ideally. This was not the case in reality. In the course of the development of the project from conceptualisation to commencement of construction, there were reservations about the viability of the project on economic, political, and environmental grounds. It was argued that the project could lose as much as US$500 million a year in operating costs (Mansley, 2003). It was also feared that the BTC pipeline would cost twice as much to operate than the two alternatives (i.e. Baku–Supsa and Baku–Novorossiysk).

On the political front, there was initial opposition to the project from Russia. It has been suggested that Russia saw the BTC pipeline as a direct threat to its own existing pipeline network (Weir, 2004). It could be argued that the reason for such opposition was fairly obvious: that the BTC pipeline would affect the country's growing monopoly over transit routes for Caspian oil. Russia insisted in 2002 that the BTC project was commercially unprofitable and that Azerbaijan did not have sufficient oil reserves to fill the pipeline, further arguing that the real objective of the project was to 'oust Russia from regions where it has historic, legitimate interests' (Socor, 2002).

Azeri energy exports and the relevance of the East–West Energy Corridor

The general argument concerning the role of the East–West Energy Corridor boils down to one key point as far as Western geo-political and geo-strategic interest in Caspian energy is concerned: that the corridor is expected to play a significant role in achieving the objective of unlocking Caspian energy reserves and securing the supply of this energy to Western markets. Winrow (2004) emphasises the strategic importance of transportation corridors to carry Caspian energy to European markets via Turkey. The environmental argument – mainly concerns about tanker traffic – also made a case for routing pipelines away from the congested Turkish Straits.

The objective of piping energy through other routes than Russia in order to ensure secure supply to Western European markets enhanced the potential role of Turkey as a corridor. Turkey's proximity to gas producers is relevant to EU energy security. Although it has been suggested that routing pipelines through Turkey was more complementary than competitive, it can be argued that Russia would view Turkey in a competitive light, rather than as a conduit.

It was Turkey's objective to create 'a corridor between countries with rich energy resources and energy consuming countries, making use of its geography and geo-strategic location', as stated by the Turkish Prime Minister in 2005. Turkey's strategy in the Caspian region included increasing investment in the oil and gas sectors of the region, enhancing mutual economic links, and securing the flow of new sources of oil and gas to Western markets. The Western view of Turkey's role in the corridor was one of dependence and security. Transit of energy through Turkey was expected to ensure a diversified supply portfolio for Europe (Acikalin, 2003).

From the Azeri perspective, Turkey's role in the corridor was obvious, and crucial. With the BTC project, and possible future oil pipeline projects, the Ceyhan terminal is vital to Azeri oil exports to Western Europe. With respect to gas, not only did Turkey's growing gas market raise the possibility of increased supply of Azeri gas, but also its developing pipeline infrastructure could serve as a new major transit system for European gas deliveries, with implications for Azerbaijan in terms of further exports to Western Europe.

Georgia's take on the corridor was one of developmental potential. According to Giorgi Chanturia (2003), then president of the Georgian International Oil Corporation, the corridor was expected to ensure the 'peaceful and secure development of the South-West European countries'. Georgia has been described as the weak link in the corridor; therefore the long-term success of the BTC and SCP projects hinges on the stability and security of the country. Concerns about the implications of reliance on Georgia's stability for the success of the transit pipelines arose following the Russian invasion of the country in 2008. While the core dynamics of the BTC and SCP pipelines have not been significantly disturbed by the invasion, questions were raised about the sustainability of future Southern Corridor projects.

Finally, one important factor in the role of Turkey in the corridor was the closeness of US–Turkish relations, which had a strong influence on the financial package for the BTC pipeline. The transportation of Caspian oil through a country then perceived as a stable ally of the North Atlantic Treaty Organization (NATO), instead of Iran or Russia, met the shared objectives of the US and Turkey: helping Turkey take the pressure off the congested Straits, and compensating Turkey for the closure of the Kirkuk–Ceyhan pipeline due to the Gulf War (Baran, 2005).

The BTC and SCP pipelines brought Turkey to the centre of energy politics. The US viewed the pipelines as significant sources of growth for the Turkish energy sector. However, it will be important to observe whether Turkish governments maintain their interest in keeping the East–West Corridor on the agenda in the long term. This will have far-reaching implications for US interest, the long-term success of the projects (and future projects), and foreign investment into Turkey.

The Shah Deniz gas field/SCP project

Since the SCP pipeline is the second part of the East–West Energy Corridor, it can be argued to be similar to the BTC project in terms of its general objectives, its suggested benefits, and the reservations over its viability. The objective of ensuring that neither Russia nor Iran developed a monopoly over pipelines from the Caspian region was furthered by this project, despite Turkey apparently being the project's chief importer.

Prior to Turkey's economic problems in 2001, the country was a rapidly growing energy importer. Between 1991 and 2010, natural gas consumption in Turkey grew from 150 billion cubic feet to 1,346 billion cubic feet (*sources*: EIA, Eurogas). From Turkey's perspective, the motivation for the Shah Deniz project is clear. From the Azeri perspective, increasing Turkish demand for gas would eventually lead to relatively easily determined long-term contracts for gas via the SCP, which would be of obvious benefit to Azerbaijan. The project at the time could therefore be justified on commercial as opposed to political grounds.

The WAGP project

ECOWAS proposed the development of a natural gas pipeline throughout West Africa as one of its key regional economic policies in 1982. Very little happened between this time and the early 1990s, when a number of studies were carried out on the commercial viability of such a project. In August 1998, a consortium of the six operator companies signed an agreement commissioning a feasibility study on the pipeline project. The conclusions of the study supported the commercial and technical feasibility of the WAGP, and projected commencement of operation as early as 2002.

Before these conclusions were arrived at, Benin, Togo, and Ghana experienced energy shortages between 1997 and 1998. These shortages renewed interest in the pipeline project. In 2002, a gas supply agreement for Ghana's Takoradi power plant was signed, with the aim of substituting gas for oil, which would significantly reduce boiler-fuel costs.

Several local environmental groups in Nigeria, Togo, and Ghana opposed the project on environmental grounds. Friends of the Earth–Ghana, for example, argued that environmental impact assessments for the project were not given sufficient priority in feasibility studies and that families in the region would be displaced as a result of the project.

A feasibility report prepared for the World Bank in the early 1990s strongly suggested that a pipeline to transport Nigerian natural gas to Benin, Togo, and Ghana was commercially viable. This conclusion was based on the gas reserves (Chevron) in Nigeria's Escravos region. The argument supporting the project's commercial viability and practicality was enhanced by the ECOWAS regional energy distribution plan of 1991 and a 1992 feasibility study on the prospects of supplying Nigerian gas to Ghanaian markets.

Given the low level of energy consumption in the West African region (especially from natural gas), it is not illogical to argue that the WAGP project was more politically driven than it was economically viable. Socio-economic factors, the low level of technology and energy intensity, as well as the generation and utilisation of electricity, especially in Benin and Togo, did not speak positively of the stage reached by this region in terms of energy demand (Omonbude, 2002). There have been criticisms of the project on these grounds – the argument being simply that there is not enough demand for the amount of gas the pipeline is able to pump. However, it appears that the project was viewed as developmental: the existence of the pipeline would create jobs, reduce energy costs, stimulate the growth of new industry from new power supplies, and so on. Ghana, for example, estimated that it would save between 15,000 and 20,000 barrels per day of crude oil by taking gas from the WAGP to run its power plants. To this end, an agreement was reached with ChevronTexaco to supply 40 million cubic feet per day of natural gas, via the pipeline, to the proposed 220-megawatt Tema power plant in Ghana.

The Chad–Cameroon oil pipeline project

The Chad–Cameroon oil pipeline project was more straightforward than the WAGP project in terms of political versus economic/commercial motives: there appeared to be a ready market for the crude oil reserves, there was a ready spot market for oil in the event of contract breaches, and Cameroon (the transit country) appeared to be fully cooperative. On the part of the government of Chad, it can be said that the primary motivation for embarking on this project was the anticipated revenue stream over a 25-year period. The Chadian population is predominantly poor, and the expectation was that the proceeds from this project would be used to increase expenditure on poverty alleviation schemes and promote economic growth.

There was equal motivation for Cameroon: the two countries (according to the World Bank's initial reports) stood to benefit from over US$1.8 billion in royalties, dividends, and taxes; more specifically, over US$500 million in transit fees were projected for Cameroon. Other anticipated benefits were infrastructural improvements in Cameroon, as well as in Chad; revenue for priority public expenditure in Cameroon; some employment generation in both countries; incentives for further oil exploration and development; and private investment in both countries.

Of interest also is the initial role of the World Bank and the IFC in this project. This can be argued to have influenced the project's progress in a number of ways, including the completion of the construction phase ahead of schedule, the relatively publicly available information on the project, and general issues of transparency. The World Bank also recognised the role of local non-governmental organisations and used them to tackle such problems as relocation and other environmental issues. As such, opposition to this project was not as fierce on any ground as it was for the other three projects discussed in this book.

The IFC played a key role in financing the Chad–Cameroon pipeline project, the objective being poverty alleviation, especially in Chad. However, Chad's failure to comply with a number of revenue management measures resulted in the World Bank's much publicised withdrawal of support for the pipeline project. Nevertheless, US interest in the project can be said to have been significant. A statement by US ambassador Linnet Deily in 2001 said that the US and Cameroon had in the past enjoyed good relations, and a strong bilateral investment treaty assured American commitment to Cameroon's further development through increased trade.

Overall, the Chad–Cameroon pipeline can be argued to be more commercial than politically driven, given the existence already of export infrastructure in Cameroon and the ease with which oil could be traded internationally.

4.3.2 The cost of the project

This section discusses the cost structure of the case study projects in terms of who bears what responsibility. The aim is to review the cost contributions of the various parties to the pipeline project. What this does, in terms of relevance to this research, is simple: it provides information on (or a numerical value for) the level of obligation of the parties to the projects, which will influence their bargaining positions (as shown in Chapter 3).

The BTC oil pipeline project

The total cost of the BTC Main Export Pipeline project (including interest on loans) was about US$3.7 billion. The cost of constructing the pipeline itself amounted to about US$2.9 billion. The key cost-related feature of this project was its international financial backing. Approximately 30% of the cost of the project was equity financed by the BTC Company participants (responsible for building and operating the pipeline, as well as ownership), and the rest was covered by third parties. The breakdown of the loans (provided at commercial rates of interest) is as follows:

- US$1 billion provided by BTC Company participants as senior sponsor lenders;
- US$250 million each in loans from the European Bank for Reconstruction and Development (EBRD) and the IFC, consisting of a direct loan of US$125 million each and another US$125 million each through a syndicate of 15 banks; and
- US$1.6 billion from export credit agencies and insurers from Japan, the US, the UK, France, Germany, and Italy.

Given the criticism levelled against the project – such as the possibility of the project losing as much as US$500 million a year in operating costs, and more importantly the possibility of the project costing up to twice as much to operate as the proposed alternative routes (Baku–Supsa and Baku–Novorossysk) – the arguably sound financial backing prepared for the project is interesting, as the political backing for the project becomes more apparent.

On the other hand, the solid financial backing for the project could be said to demonstrate the confidence of the financiers in the project's commercial viability. Ideally, private lenders would not sponsor a project unless they were convinced that the anticipated returns justified the costs. The BTC Company participants have tried to reinforce this point. According to them, the financing agreements came about after more than two years of extensive scrutiny of the project's environmental and social effects, in addition to a public consultation process. In addition, the loans are subject to formal quarterly audits by the lenders and are conditional on the project continuously meeting public commitments. The involvement of multinational export credit agencies has also enhanced the transparency of the project.

The cost implications for the countries along the route of the pipeline can be deduced from the analysis above. In the case of Azerbaijan, the producing country, its main objective is to get its oil to Western markets. Azerbaijan expects gains in oil revenue over a 20-year period from the development of the offshore ACG oil fields. With SOCAR's 25% stake in BTC Company (Table 4.2), the Azeri government bears direct responsibility for ensuring the successful construction and operation of the pipeline; however, since its share in overall project cost responsibilities is up to 30%, Azerbaijan does not bear the full cost burden of the project alone. Therefore there is a limit to the extent to which it can exercise control over the project.

Georgia, the transit country, is notably not a BTC Company participant, and there is no record of Georgian influence in the rest of the financial package prepared for this project. Georgia has no off-take agreement during the operation of the pipeline. It does not contribute financially to the construction, operation, or maintenance of the pipeline. It was originally expected to earn almost US$600 million in transit fees, demonstrating the characteristics of a pure transit country as discussed in section 3.4. On the basis of the obsolescing bargain, Georgia assumes an advantageous bargaining position when the pipeline commences operation; it is in a position to seek increased rent in transit fees, for example. Since it has been defined (for the purposes of this analysis) as a pure transit country, Georgia would not fundamentally stand to lose out should such demands for an increase in transit fees force the operation of the project to cease; it could theoretically simply revert to its position prior to receiving earnings from the operation of the pipeline.

The Turkish national oil company (Turkiye Petrolleri Anonim Ortakligi) is a BTC Company equity participant, and thus Turkey bears some of the costs of the project and contributes to the construction, operation, and maintenance of the pipeline. It is not the final destination of the transported crude oil, so Turkey also plays the role of a transit country.

The Shah Deniz gas field/SCP project

The total cost of the Shah Deniz gas project and the SCP was estimated at US$2.7 billion. Total investment for the project was said to have reached US$3.2 billion, after including financing costs and interest during the construction period. The pipeline itself cost around US$1 billion to construct. The financial package for the project included an EBRD loan to SOCAR of about US$250 million; the state company met the other US$70 million investment obligations to the project.

As with the BTC project, Georgia does not contribute to the SCP directly in terms of project costs. It is not a participant in the SCP's ownership, construction, or operation. Georgia has, however, negotiated a gas off-take agreement. It will get 5% of the gas transported through the pipeline as its transit fee, with an option to purchase an additional 0.5 billion cubic metres per annum.

Azerbaijan and Turkey were directly involved in financing the project, as reflected in their equity participation (Table 4.3). Azerbaijan's broad objectives in this regard were the development of the Shah Deniz gas fields and the securing of long-term contracts for its piped gas. The Turkish objective was to secure sources of gas for its distribution network, and the SCP project met this requirement.

The WAGP project

The total investment costs for the WAGP project were originally set at US$500 million. However, these costs more than doubled to over US$1 billion. The project was financed by shareholders of the West African Gas Pipeline Company (WAPCo) on the basis of a gas purchase contract with the Volta River Authority of Ghana as one of its foundation customers, accounting for about 90% of the demand for gas to be supplied initially by the project (*source*: World Bank). Communauté Electrique du Bénin of Benin and Togo would account for the remaining 10% of the demand.

As with the BTC project, the interesting aspect of the cost and financing of the WAGP is its financial backing. Nigeria was keen to ensure the success of the project, and in July 2004 it approved a loan facility to Ghana (US$40 million) to help Ghana fulfil its part in financing the project. The primary buyer for the gas is the Ghanaian power sector.

As Table 4.4 shows, there is some form of equity participation in WAPCo by the four countries involved in the WAGP project. The implication of this involvement is straightforward: each country has some form of cost obligation and, given this, it would appear to be in the best interests of the transit countries to ensure uninterrupted transit.

The Chad–Cameroon oil pipeline project

The cost of the Chad–Cameroon oil pipeline project (field development and construction of the pipeline) amounted to about US$3.5 billion. According to the EIA, the financial package agreed for the project breaks down as follows:

- Cameroon Oil Transportation Company (COTCO, a joint venture between a foreign consortium and the governments of Cameroon and Chad) provided 59.2% of the total investment (US$2.2 billion).
- Commercial banks and export credit agencies contributed 16.1% (US$600 million).
- Capital markets provided 10.7% (US$400 million).
- The IFC issued a US$100 million loan to COTCO, and an additional US$300 million was provided through commercial banks.
- The World Bank provided loans of US$53.4 million for Cameroon and US$39.5 million for Chad, for the purpose of funding the two governments' stakes in the project.

In terms of meeting the cost obligations, responsibility is shared between the two countries, which can be argued to imply commitment to the success of the project. This has further implications for the justification for, determination of, and adherence to the transit fee which Cameroon will earn.

4.3.3 The economic and geo-political relevance of the project to the transit country

This section analyses the economic value of the four projects to the transit countries.

The BTC oil pipeline project

In terms of economic relevance, the importance of the BTC project to Georgia is obvious. Georgia gains the transit fee revenues from the BTC pipeline.

On the assumption that the Baku–Supsa pipeline would continue at maximum capacity and the BTC pipeline would be used for exports of ACG oil that could not be accommodated by the Baku–Supsa pipeline, and on the basis that the transit fee for the Baku–Supsa pipeline exceeded that for the BTC pipeline, Georgia's transit revenues would have reduced if oil shipped via the Baku–Supsa pipeline had been redirected through the BTC pipeline (Billmeier et al., 2004). This notwithstanding, Georgia stood to make a significant amount of revenue from the project, which gave the BTC pipeline a high degree of relevance to Georgia on commercial grounds.

Georgia has been identified as a pure transit country as far as this project is concerned. In terms of cost responsibilities, it makes no financial contribution to the construction, operation, and maintenance of the pipeline, although it can be argued that there are socio-cultural and environmental costs to Georgia resulting from the construction of the pipeline through its territory. In strict cost-of-project terms, therefore, the importance of the project to Georgia is the rent it gets from transit fees. This is in addition to the jobs that will be created in the region, the proposed patronage of local companies for supplies for maintenance of the pipeline, and so on. It can, therefore, be argued that Georgia would not be worse off if the pipeline did not pass through its territory. This has obvious bargaining implications for the parties to the project should Georgia seek more rent from the project via increased transit fees. However, given the strategic alliances with Turkey and the keen interest of the US in this element of the East–West Corridor, it does not appear likely that Georgia will disrupt the pipeline, although the political climate in the country can be said to be unstable.

Turkey was projected to earn up to US$3.4 billion from its involvement in pipeline and terminal operations, transit fees, and upstream investments (*Petroleum Economist*, 2005). The BTC project is economically (and geo-politically) relevant to Turkey for three key reasons:

▸ After the Gulf War, the country needed to recover losses suffered as a result of the closure of the Kirkuk–Yumurtalik pipeline transporting Iraqi oil to terminals in Ceyhan.
▸ There was a realisation that the potential value of central Asian and Caspian oil resources would be greatly enhanced with Western consumer access.
▸ As a NATO ally and strategic partner of the US, Turkey believed it was best suited to reap the rewards of such political leverage. This would benefit Turkey in terms of its objective of further institutional integration on a regional and international scale. This can be said to be one major reason for the country's direct involvement in the project. On the basis of its direct involvement in meeting cost obligations for the project, it can be argued that there is more for Turkey to gain than transit fees from the BTC.

The Shah Deniz gas field/SCP project

Georgia receives 5% of the gas transported through the SCP as a fee, and it is permitted to sell on that gas. The transit fee earnings from the SCP for Georgia are, however, dependent on the ability of its domestic market to pay for the gas. According to Billmeier *et al.* (2004), there are three categories of consumer in Georgia's domestic gas market: large power-generation/industrial clients, distributors, and mainly small-scale businesses. Payment discipline is highest in the small-scale segment; the large-scale segment is less disciplined because the organisations in this category are deemed to be of strategic value to the Georgian economy.

The piped gas contributes 1% of Georgian gross domestic product (GDP) in additional revenues if the pipelines are used at full capacity. Given Georgia's relatively low tax revenue, this income can make a significant contribution to the government's resource base, but not of the same magnitude as would revenues for a producing country.

For Georgia, the SCP is different from the BTC project in that the country receives a percentage of the gas transported through the pipeline as off-take. It was expected to benefit from the SCP project in terms of the opportunity to diversify its sources of gas. Shah Deniz gas would also provide Georgia with an opportunity to ease its gas feedstock problem. It is estimated that the volumes of gas supplied from the SCP would be sufficient to generate about 20% of Georgia's winter peak load energy needs. This translates to about 80% of Tbilisi's peak load needs. It was also expected that the project would kick-start the development of a competitive gas market, which could potentially result in reduced prices in the long term.

In the context of Georgia's previous definition in this chapter in relation to the BTC pipeline as a pure transit country, the case is different with the SCP. This is simply because of the dependency of the domestic Georgian gas market on Shah Deniz gas. The project can therefore be cautiously suggested as being of more value to Georgia than the BTC pipeline.

The WAGP project

Benin and Togo receive a percentage of the gas from the WAGP project as off-take for their respective electricity generation needs. It is argued that the WAGP is essential to the development of the electricity sectors of both countries. Power distribution utilities have been privatised (Togo's in 2000; Benin's in 2006), arguably in anticipation of the opening up of the power generation and distribution markets in both countries.

Benin and Togo have fairly significant power generation sources (thermal and hydro). These have, however, been argued to be expensive, and they have been problematic over time. It was said that the WAGP would provide a cheaper alternative. The World Bank estimated that Benin and Togo (and Ghana) could save as much as US$500 million in energy costs over a 20-year period by substituting WAGP gas for more expensive power generation fuels.

The Chad–Cameroon oil pipeline project

There were two important aspects of the pipeline for Cameroon: financial and developmental. First is the financial benefit. According to the EIA, the original financial terms showed that Cameroon stood to earn about 46 cents (US$0.46) on every barrel of oil transported through the pipeline. COTCO stood to gain from the project, especially in terms of fiscal advantages (Ndum, 2004); it was expected to benefit from a number of tax exemptions, which included exemptions on a special petroleum products tax and turnover tax. In terms of benefit to the state, both the World Bank and COTCO envisaged that Cameroon would earn US$500 million in transit fees and other taxes over the 25-year life of the project from the projected 900 million barrels of oil that would pass through the country (Ndum, 2004). COTCO would then pay the transit fee to the Republic of Cameroon in consideration of concessions granted to the company. Of note at this point is that the transit fee was not designed to be responsive to fluctuations in oil prices, the implication being that Cameroon would not benefit from increased transit revenues as a result of an upturn in international oil prices.

Second is the developmental benefit of the pipeline. The Chad–Cameroon pipeline is located near the Logoni Basin. What this means for Cameroon is that future petroleum production at the basin could be significantly facilitated by the presence of the pipeline. For definitional purposes, then, Cameroon is not a pure transit country as far as this project is concerned, given its direct involvement in the project and the potential benefits both to the state and to COTCO.

In summary, although the BTC pipeline project provides substantial revenue for Georgia, it is still a pure transit country by definition, whereas Turkey's direct involvement in the project implies that the BTC pipeline has greater significance for the country. Georgia's likelihood of renegotiating transit terms will, however, depend on the state of play with respect to the strategic alliances of the government of the day. The case is different for the gas projects reviewed here (the SCP and the WAGP); these projects are more significant to the transit countries. Georgia, Benin, and Togo as transit countries need the gas to meet domestic demand. Strategic alliances also come to play in this regard, especially in the WAGP case. In the case of the Chad–Cameroon pipeline project, Cameroon is directly involved in the project, and there are potential developmental advantages to the country from the pipeline, both directly and (especially) indirectly. There is, however, a significant potential difficulty in the Cameroonian case because of the unresponsiveness of the transit fee to oil prices. This manifested, for example, in the 2009 claims for a renegotiation of the transit terms by the Cameroon government.

4.3.4 The value of the transit route to the entire project

The BTC oil pipeline project

The importance of the Georgia–Turkey route to Azerbaijan's oil export objectives is quite significant, as well as obvious. However, given the level of criticism of the choice of the BTC pipeline route through Georgia and Turkey, it is pertinent to examine the value of this route to the project in an economic and political context.

There are three feasible export routes for Caspian oil: Russia, Iran, and Turkey (i.e. Georgia–Turkey). The question of whether the chosen route for the pipeline was the cheapest alternative has been discussed already in this chapter, and there are basically two opposing answers: sustained criticism of it as an alternative versus the political and commercial backing it has received. For example, it was argued that the project was marginal in terms of financial return, which, at a rate of 10.9%, is less than BP's target rate of return on capital (Mansley, 2003).

The NREP (Baku–Novorossiysk) and WREP (Baku–Supsa) could be argued to be existing alternatives to the BTC pipeline (in terms of the primary Azeri objective of exporting crude oil from the Azeri–Chirag fields), and could also be argued to be cheaper alternatives. For example, in 1998 the Azerbaijan International Operating Company estimated the costs of the BTC and Baku–Supsa pipelines at US$3.7 billion and US$1.8 billion, respectively. Whereas the BTC pipeline was expected to raise the cost of a barrel of oil by US$4, the Baku–Supsa pipeline was expected to raise it by US$2 (*MEES*, 1998). However, the Baku–Supsa and Baku–Novorossiysk pipelines both terminate at export terminals for further transport via the Black Sea. The Turkish Straits congestion argument implies (especially for the key proponents of the BTC project) that given the demand for increased Azeri crude exports to Western markets, there was justification for the route through Tbilisi.

Given this apparent demand for more pipelined crude oil from Azerbaijan, the question of whether Georgia–Turkey is a cheaper alternative than a route via Russia or Iran comes to the fore. Russia is the least-cost option, given its extensive existing pipeline network and relative accessibility to European buyers. Iran is the next best economic alternative. This makes the Georgia–Turkey route the most expensive option. The choice of either of the other two options, however, would have geo-political implications. Although Russia is the least-cost solution, another pipeline through Russia to Western Europe would be seen as giving it too much control, which can be argued to have serious security of supply consequences. And poor relations between the US and Iran do not encourage the Iran option. The next best alternative to the Russian or Iranian option is the BTC Georgia–Turkey route.

The Shah Deniz gas field/SCP project

One key point to be considered in the case of the SCP is that it is a gas project, and so differs from an oil pipeline in certain respects. Gas sales are tied to specific markets and are supported by sales and purchasing agreements; for oil, the markets are more open. In analysing the value of Georgia as the transit country for the SCP, therefore, it is pertinent to consider the choice of target market, the alternative methods of transporting Shah Deniz gas, and the other possible routes, from an economic and political perspective.

The reasons for choosing Turkey as the export country included its ability to pay for the gas, its existing domestic market, and Georgia's support and willingness to be a transit country. The other export options (especially Russia, Iran, and Armenia) were not acceptable for the Shah Deniz project, for reasons of economic viability and political constraints. Although there are already gas pipeline networks across Russia, exporting Shah Deniz gas via this route would go against the Western (US) objective of stemming the Russian monopoly over Caspian energy exports to Western Europe, as discussed throughout this chapter. Armenia was not a politically acceptable route for three key reasons:

- the unresolved military conflict between Armenia and Azerbaijan over rights to Nagorno-Karabakh;
- the close ethnic ties between Azerbaijan and Turkey, which is why it closed its border with Armenia; and
- political tensions between Armenia and Turkey.

After the destination market for Shah Deniz gas had been chosen, transportation alternatives were considered: a pipeline to Turkey, gas by wire (i.e. conversion of gas to electricity and transmission via high-voltage lines), and LNG export via sea, rail, or road tanker. After careful consideration, it was decided that the pipeline option was the most economically viable. The project would cover a long distance, making the gas by wire method expensive when the energy losses associated with power transmission are taken into account. LNG export was rejected because of the large number of trips that would be required, as well as other potential problems, such as emissions, likely loss of inventory, and safety and environmental risks (e.g. the risk of tanker collision).

The decision to build a pipeline to Turkey required the project planners to select the most feasible route for the SCP. The possible transit routes were Iran, Armenia, and Georgia. The reason for not selecting Iran as the transit route was largely political. As with the BTC project, the continued sanctions on Iran, as well as the involvement of US investors and partners in the Shah Deniz PSA and SCP project, prevented the choice of Iran as the transit country. The case was similar for the Armenian alternative. Azerbaijan and Armenia had had serious political disputes in the past, and the wishes of the Azeri government had to be respected. This implied that the next best alternative – the only other choice for that matter – was through Georgia.

The question of whether Georgia was the cheapest route, therefore, does not have much of a role to play in this analysis, given the unfeasibility of the other options. If the project was to be embarked upon, Georgia had to be the transit route. This (as will be shown later in the chapter) has implications for the bargaining power of Georgia.

The WAGP project

Ghana is the primary market for WAGP gas. The piped gas is expected to serve the Ghanaian power sector. Further destinations are also being considered, such as an extension of the pipeline to the Ivory Coast. Taking into consideration such factors as distance and LNG versus pipeline economics, piping the gas to Ghana can be argued to be the cheapest method of exporting Nigerian gas, compared with gas by wire and LNG transportation.

The question of the choice of the transit countries (Benin and Togo) is straightforward: this is the most economical route for the pipeline. There is no practicable alternative to this offshore transit route. Without the cooperation of these two countries, therefore, the WAGP project will not succeed long term, so the role of Benin and Togo as transit countries is pivotal to the success of the entire WAGP project. As already discussed, Benin and Togo are to receive a percentage of the transported gas in off-take volumes, and in addition they have equity responsibilities to WAPCo, however minimal. They have experienced difficulties developing their power sectors, and WAGP gas is vital to their long-term power sector development projects. As is shown later in this chapter, this has implications for the bargaining power of Benin and of Togo.

The Chad–Cameroon oil pipeline project

The role of Cameroon as a transit country in the Chad–Cameroon oil pipeline project is similar to that of Benin and Togo in the WAGP project. It can be argued that without cooperation from Cameroon, the pipeline project simply would not work, because Cameroon was the only feasible transit route for Chadian oil to get to offshore loading facilities. Transit through Nigeria and the Central African Republic would have been long and expensive. It is important at this point to reiterate the significance of the pipeline project to Chad. The involvement of Cameroon in the project was, therefore, crucial to its success. Cameroon also stood to benefit from the pipeline running through its territory, as the pipeline enhanced the developmental potential for its Logoni Basin fields, which are close to the route for the pipeline.

4.3.5 The economic and political environment

Azerbaijan, Georgia, and Turkey

According to authoritative sources on the political environment, countries in the Caspian region have been dominated by political instability. Political life in the region has been typified by disputes over territorial integrity, ethnic clashes, and instability of governments. Azerbaijan, Georgia, and Turkey are members of a range of international and regional organisations, including the World Bank and the International Monetary Fund (IMF). A deeper relationship with Europe is promoted by some of these organisations, while others promote trade and regional stability, the development of market-oriented economies, and democratic institutions. The UN has played a significant role in the region, acting as peace-keeper and arbitrator in a number of internal and international disputes.

The main export product for Azerbaijan is oil, production of which declined until 1997, but has been on the increase since then. Through negotiated PSAs with foreign companies, the long-term goal for Azerbaijan is to generate funds for future industrial development. Azerbaijan has experienced significant difficulties (as have other former Soviet republics) in transiting from a command to a market economy, but its economic prospects have been enhanced by its vast energy resources. There have also been other significant constraints on the Azeri economy. One major obstacle to the economic progress of the country lies in the continuous conflict with Armenia over the Nagorno-Karabakh region. Another constraint is corruption. According to Transparency International, Azerbaijan's Corruption Perception Index ranking varied between 158th and 134th between 2008 and 2010 (Table 4.5).[1]

TABLE 4.5 *Corruption index for selected countries, 2010*

Country rank	Country	Corruption Perception Index 2010 score	Surveys used
56	Turkey	4.4	7
62	Ghana	4.1	7
68	Georgia	3.8	7
134	Nigeria	2.4	7
134	Azerbaijan	2.4	7
146	Cameroon	2.2	7

Source: Transparency International website.

An important aspect of the economic direction of Azerbaijan is the change in its flow of trade. Trade with Russia and other former Soviet republics declined with the end of the Soviet Union and was replaced with increasing trade with Turkey and Western European countries.

In terms of politics, the absence of political and legal stability in the region has a long history. Azerbaijan was the first Soviet republic outside the Baltics to declare national sovereignty. It gained formal independence in 1991 and agreed to a new constitution in 1995. Problems have, however, ensued from its transition to an independent state. Azerbaijan has experienced some economic disruptions since its independence, coupled with such political problems as armed struggle over border disputes; thus, the Azeri political situation can be said to be quite volatile, even during long stretches of peace.

The Republic of Georgia imports the bulk of its energy requirements, including natural gas and oil products. Its only sizeable domestic energy resource is hydropower. The Georgian economy has experienced severe disruptions as a result of civil strife. However, the IMF and the World Bank have assisted the growth of the Georgian economy in one way or another since 1995. These efforts have resulted in positive GDP growth and a significant check on the rate of inflation. However, Georgia carried a burden of limited resources owing to such factors as a poor tax administration system (outright failure to collect tax revenues) and energy shortages. The Tbilisi distribution network was privatised in 1998, but the venture proved unprofitable because of low collection rates. Like Azerbaijan, Georgia is beset by high corruption levels (Table 4.5). This has affected the economic development of the country and contributed to existing political tensions in the region.

One important aspect of the Georgian economy is that its long-term economic growth prospects currently hinge on its role as a transit route for pipeline projects.

The Republic of Georgia has also experienced political difficulties since gaining independence in 1991. Over the centuries, Georgia was the object of great rivalry between Persia, Russia, and Turkey. In recent times the US has shown major interest in the country's security and stability. The US has provided training and support for the Georgian army. The close ties between the two countries are being closely watched by Russia. Georgia is one of the poorest former Soviet Union countries and still depends on Russia for its energy supply. The political climate in Georgia has been dominated by periods of civil war and unrest. The independence aspirations of the breakaway regions of Abkhazia and South Ossetia have in the past led to violence (e.g. in the Russian invasion of 2008), and despite the relative stability built by diplomatic efforts in recent years, there are still tensions over both regions.

Turkey has recently experienced erratic economic growth and heavy imbalances. Its growth was interrupted by sharp declines in output in 1994, 1999, and 2001, and the country had high inflation rates. Foreign direct investment (FDI) has been relatively low, and in the early 21st century high trade deficits and serious banking sector weaknesses forced the economy into crisis, pushing Turkey into recession. The IMF has provided significant financial support to improve the situation. This assistance is reflected in the country's tightening of fiscal policy.

Most of Turkey's recent politics (and arguably its economics) was shaped around its plans to accede to the EU. It is a NATO member, implying an obvious Western outlook in the country's international diplomatic efforts. Turkey initially made significant progress towards meeting the political, economic, and institutional requirements for EU accession. However, the country's political climate has been beset by years of conflict, leading to mass evacuations from the conflict regions, with severe economic consequences. This is argued to have seriously affected its chances of EU accession, which is notably opposed by EU leader countries such as France.

Nigeria, Benin, Togo, and Ghana

The Nigerian economy is heavily dependent on the oil sector, which contributes more than 90% of export revenues, more than 70% of government revenues, and roughly one-third of GDP. Nigeria has also started to fully exploit its natural gas reserves and to harness previously wasted gas from oil production. The two main gas projects are an LNG project and the WAGP. Despite the economic growth potential of the country, a significant percentage of the population falls below the poverty line. Basic social indicators place Nigeria among the poorest countries in the world, even though Nigeria accounts for about half of West Africa's population and more than 40% of the region's GDP.

Despite Nigeria's recognition of the effects of over-reliance on the oil industry, the country's future economic prospects are still heavily predicated on revenue from its oil and gas resources. The level of corruption is also quite high (Table 4.5), which has had negative political consequences over the years.

In terms of politics, the country currently has a democratically elected government after 28 years of military control following its independence from British rule. Given the country's cultural diversity (more than 250 ethnic groups), it has been difficult for Nigeria to strike a political balance that sufficiently takes the interests of these diverse groups into consideration. In addition to corruption and the violence resulting from ethnic and religious clashes, there has been significant unrest in the oil-producing communities; thus, the Nigerian political climate can be said to be rather tense. In regional terms, however, Nigeria plays a leadership role in ECOWAS and the African Union.

The economy of the Republic of Benin relies heavily on subsistence agriculture, cotton production, and regional trade (especially between Cotonou and Lagos). Annual GDP growth has been modest at best, hovering around an average of 4%, but this growth has been counter-balanced by a rapid rise in population. In terms of its economic outlook, the country has pursued a privatisation exercise, as well as a foreign investment drive, with the objective of boosting its electricity, telecommunications, and agricultural sectors. The country expects to make long-term gains from the WAGP project through gas off-take in return for its transit role in the pipeline.

There have been improvements in Togo's agricultural sector, services sector, and economic policy. Togo expects to boost its electricity sector with the gas off-take earnings from its role as a transit country in the WAGP project. Further economic development in the country is to some extent dependent on political stability.

In terms of per capita output, the Ghanaian economy is twice as wealthy as most of the poorer countries in the West African region. It is, however, heavily reliant on international financial and technical assistance. Ghana's domestic economy revolves around subsistence agriculture (which accounts for more than one-third of GDP). Mining (gold) and oil are also key industries to the economy. The economy experienced difficulty towards the end of the 20th century, when the prices of its main exports (gold and cocoa) dropped significantly, and prices of petroleum imports increased. The country recently carried out macroeconomic reforms. In the energy sector, for example, an automatic adjustment formula for retail petroleum prices was adopted following the near-doubling of petroleum prices in early 2003. Electricity tariffs were also raised by 72% between 2002 and 2003. Projects such as the WAGP are expected to bring secure energy sources for Ghanaian power projects, which in turn will enhance economic development in the country. Vast offshore oil discoveries have also necessitated a reform of the petroleum upstream sector.

In terms of politics, the climate in Ghana is relatively stable compared with the rest of the West African region. The country has enjoyed a sustained period of democratic rule and is one of the less corrupt states in Africa (as Table 4.5 shows). There is not significant documentation of disputes in Ghana, either domestically or internationally.

Chad and Cameroon

Chad is one of the world's 10 poorest countries, according to the World Bank. In the circumstances, however, Chad has managed and maintained a satisfactory macroeconomic track record since 1994. Its real annual GDP growth rate averaged just over 5% between 1994 and 2003, and has averaged more than 9% since 2001 as a result of oil-related investments. The Chad–Cameroon pipeline is one of the largest projects the country has embarked upon. Construction was completed a year ahead of schedule, apparently reflecting the significance of the pipeline to the country's economic ambitions. There has been significant assistance from the World Bank for the country's social sector, transport and energy, agriculture, and capacity-building.

In terms of politics, Chad is an ethnically heterogeneous state that appears to have been founded without much consideration of the dissimilar tribal cultures or religions and economic realities. There are about 200 ethnic groups, which, as in the Nigerian case, could prove a source of conflict in the country. However, there has been a certain degree of stability in the recent past in the country.

Cameroon is a mainly agricultural economy. Petroleum products make up more than half of its major exports. A sharp decline in the prices of the country's major exports in the mid-1980s severely affected the Cameroonian economy. It also suffered a fiscal crisis during this period, in addition to years of mismanagement and corruption (Table 4.5). The economic challenges for Cameroon include promoting a stable macroeconomic framework, economic diversification, private sector development, and sustainable natural resources development. The World Bank has also been closely involved in providing financial assistance for the restructuring of the country's economy.

Cameroon has enjoyed relative political stability compared with most countries in the region – and indeed in Africa. On the international front, however, there have been major border disputes with Nigeria over the Bakassi Peninsula, on which the International Court of Justice had to arbitrate. Despite the Court's ruling (which favours Cameroon), there have been skirmishes in the area recently.

This section set out to analyse economic and geo-political influences on the pipelines. In the case of the BTC and SCP projects, border disputes and strategic alliances (in some cases arising from such disputes, e.g. Azeri–Turkey relations) were a major determinant of the choice of transit route. With respect to the WAGP, economic development goals can be deduced to have been the key factor driving the proponents of the pipeline project to ensure its feasibility. Although there is an immediate need for the exported gas in the Ghanaian power sector, commercially feasible gas needs are not immediately evident in the transit countries. With respect to the Chad–Cameroon pipeline, the Chadian economy could certainly benefit from the project's revenue streams. The economic influences on (and indeed motivation for) the project are obvious. The same can be said for Cameroon, although its economy is in a significantly better state than Chad's.

4.3.6 Factors relevant to the bargaining process

The characteristics outlined in the sections above all have implications for the bargaining positions of the participants in the case study pipeline projects. As suggested throughout this chapter, these bargaining positions are influenced significantly by a mix of economic and political factors, which might change during the life of the project. I argue that the bargaining positions of the countries involved will also differ (however slightly) according to the hydrocarbon being exported. This section identifies factors from the case studies that will be used in section 4.4, which applies the bargaining principles discussed in Chapter 3.

Oil pipelines

With the BTC project, Azerbaijan's primary objective is to get its oil to Western markets. It already does so using the NREP and WREP to the Black Sea. Since the case for the commercial viability of the BTC pipeline was made by the participants and financiers (with reference to other arguments regarding the potential choking of the Turkish Straits), it is useful to conclude that the pipeline is also essential to the development of the oil fields in Azerbaijan, and that there are sufficient reserves to meet demands. The revenue from these projects can also be deduced to be of great importance to the Azeri government. From this perspective, therefore, Azerbaijan can be argued to be in a weak position in terms of its bargaining power, because of its established need to export its oil. However, the transit countries have also been shown to have interests in the pipeline.

The Georgian case is notable: it is, to all intents and purposes, a pure transit country as far as the BTC project is concerned, but strategic interests and alliances (economic and geo-political) seem to check its bargaining power. From the discussions in Chapter 2, we know that the possibility of a pure transit country exercising its bargaining power (power gained once the line is operational) exists. It is argued here that a scenario in which there was no pipeline at all would not make Georgia worse off; on that basis (with the exclusion of environmental costs), it can exercise its bargaining power in transit fee negotiations or renegotiations. However, Georgia can also be argued to be better off with the pipeline passing through its territory, from transit fee revenue alone.

Turkey realises its role not only as a transit country, but also as the key to the US foreign policy driven East–West Energy Corridor. It is, however, also aware of this role in the context of the benefits it will accrue, both commercial and political. Turkey's bargaining power is greatly influenced by the potential benefits of having the pipeline pass through its territory to the Ceyhan terminal, where the oil can be traded. Another benefit to Turkey from the BTC project is the revenue it will earn from transit fees.

The question of whether Turkey will exercise (or has any) bargaining power over the project (either in collusion with Georgia or alone) can be answered by reference to three key political factors:

▸ The involvement of the US ensures that a watchful eye is kept on the regulation of the project.

- Georgia has been promoted (again by the US) as vital to the development of the East–West Corridor; as an ally, it is not likely to benefit immensely from disrupting the project.
- In terms of Turkey's relations with the EU, any perceived 'negative' role in the BTC process could threaten its position, although the situation could change significantly if Turkey loses interest in, or support for, EU accession.

In the case of the Chad–Cameroon pipeline project, the situation of Chad appears slightly more desperate than that of Azerbaijan. Chad's economy should benefit significantly from the development of its oil fields and revenue accrued from the sale of its crude oil. In addition, there is no commercially feasible alternative to the route through Cameroonian territory. From this perspective, it can be argued that Cameroon could exercise its bargaining power in the negotiation and determination of the transit fee.

On the other hand, Cameroon – despite being, for the purposes of this study, a pure transit country, which could therefore exercise its bargaining power – will be better off with the pipeline passing through its territory for two major reasons: transit fee revenue and the fact that the pipeline runs close to the Logoni Basin, which brings significant development prospects for the oil fields in the basin. The problem of transporting oil from the Logoni Basin to Kribi for marine export could be solved in the future by connecting to the Chad–Cameroon pipeline (of course taking pipeline technicalities such as throughput and capacity into consideration).

There is, however, one problem with the potential transit fee revenue in the Cameroonian case: as already noted in this chapter, the transit fee per barrel of oil transported via the pipeline does not change with fluctuations in the oil price. Thus, Cameroon might not be entirely comfortable with the rent it gets from the fee, and might be fully aware of the potential to squeeze more rent when the oil price is high, as was the case during the 2003–2008 global oil price rise.

At the beginning of this section, it was mentioned that the bargaining positions of the countries involved will differ according to whether the pipeline carries oil or gas. In Chapter 2 the technical differences between oil and gas pipelines were discussed. In addition to differences in costs and pumping requirements, there are two major factors to consider in the assessment of the bargaining power a transit country can exert in gas pipeline projects:

- *The off-take.* This is more common in gas pipeline projects than in oil pipeline projects. The transit country could be paid an off-take

of the gas transported for its domestic use (power projects, heating, etc.) as opposed to a transit fee, or it could be paid both.
- *The nature of gas pipeline contracts and the market structure.* Owing to the nature of gas pipeline contracts (which usually last 15–40 years), there is a close contractual link between the producers and the consumers. Any disruption of the project results in serious devaluation of the investment.

Thus, there are differences in the factors to be considered when we analyse the bargaining positions of the countries involved in gas pipeline transit.

Gas pipelines

Georgia does not own or operate any part of the Shah Deniz pipeline, it is not an equity participant in the company running the project, and there is no commercially feasible alternative to the Georgia route. Turkish involvement in the project also implies significant demand for SCP gas. Georgia, therefore, seems to have the bargaining advantage. However, Georgia gets an off-take of SCP gas, and this changes its bargaining position significantly for two reasons. First, it has a gas feedstock problem; therefore, it will benefit from this off-take in meeting domestic gas requirements (especially during winter peak load periods). Second, the gas off-take arrangement (with an option to buy more) will ease the power shortages it has experienced over time. On these grounds, Georgia cannot afford to interfere with the project by exerting its bargaining power, as such interference would not only affect the project in its entirety, but would also be detrimental to its domestic gas supply.

Another factor is US influence (just as with the BTC project, and possibly future projects from the Caspian region). The SCP constitutes the second part of the East–West Corridor, and the implications of US concern over the success of the project for the role of Georgia as a transit country are obvious.

The WAGP case is peculiar. It appears that the transit countries (Benin and Togo) have been compelled to accept off-take gas from the pipeline. Given the character of the project – there is no feasible alternative route; Nigeria must export its gas; Ghana needs the gas to enhance its power projects; Ghana and Nigeria have entered into long-term sales agreements; Benin and Togo are not entirely politically stable; both transit countries contribute very little to the project; and there is not much demand for gas in these transit states (given their stage of energy transition) – one could easily expect Benin and Togo to fully exert their bargaining power in the process. However, a number of factors could significantly affect the way they use their roles. First, there are power projects (some in the planning stage, others in the construction phase) that will benefit from WAGP gas. Second, these countries have no feasible alternative sources of gas, in the region at least. Third, there is a power supply problem in the urban regions of both countries. Parts of Benin already rely on electricity supply from Nigeria. In addition, the proposed West African Power Pool project will utilise WAGP gas. Fourth, political relations between the countries have an impact. Given the role of Nigeria and Ghana in fashioning West African economics and politics, it does not appear to be in the best interests of Benin and Togo to disrupt the pipeline.

4.4 Implications of the principles of bargaining theory for the case studies

The principles of bargaining theory applicable to transit pipelines were discussed in Chapter 3. Section 4.3.6 highlighted the factors that would influence the bargaining positions of the parties to the case study transit pipeline agreements. This section looks in more detail at the implications of the bargaining positions of the parties involved (especially the transit countries) in the context of the bargaining principles. It combines the preliminary analysis of relevant factors in section 4.3.6 with the bargaining principles set out in Chapter 3, drawing on evidence from the case studies as analysed in previous sections of this chapter.

4.4.1 The common trade interest principle: bargaining implications

It was established in Chapter 3 that there will be no bargaining situation without a common interest to trade, and that the parties will have conflicting interests over the price at which to trade. The key question here relates to whether the project is attractive enough to generate and maintain interest especially for the transit country to enter into, and honour, an agreement with the producer country. Other important issues are the relevance of the project to the producer country, the relevance of the transit fee or off-take to the transit country, and a level of outside interest (or outside influences) on the project's existence and continuous operation.

DOI: 10.1057/9781137274526

From the case studies, there is obvious common interest to trade. In the BTC project, for example, the value of the pipeline to Azerbaijan is significant. Increased demand for Azeri crude implies, *ceteris paribus*, more revenue to the Azeri government from the pipeline. Of more significance to this analysis, however, is the value of the project to Georgia and Turkey as transit routes, given their potential bargaining power on the basis of the obsolescing bargain principle. Georgia's continued dependence on Russia for its energy supply can be argued to be another factor that could influence the transit country's behaviour in terms of contract renegotiations. Similar cases can be argued for the parties to the Shah Deniz project, the WAGP, and the Chad–Cameroon pipeline.

It cannot, however, be concluded that these factors will keep bargaining power with the transit countries. A number of factors have been identified in the cases which could have serious positive influence on the bargaining power of the transit countries. These are classified under three headings: *economic benefit to the parties involved, ownership/ equity participation in the project*, and *external influence from international organisations*.

Economic benefit to the parties involved

In the BTC and Shah Deniz cases, Georgia has based its long-term economic growth prospects on the revenues to be made from its role as a transit route for the pipelines. As already discussed in this chapter, FDI in the Turkish economy is low. In the case of the WAGP, Benin stands to benefit from the gas pipeline for its power needs and also the development of its telecoms and agricultural sectors. Togo also expects to boost its electricity sector using gas off-take from the pipeline. More than half of Cameroon's export revenue is petroleum based, hence the importance of the Chad–Cameroon pipeline to the country.

Ownership/equity participation in the project

As with most modern transit pipeline agreements, the four pipeline cases in this research show some level of equity participation by each of the parties to the project, however unequal (although Georgia is not a BTC Company participant). This clearly implies each party has a vested interest in ensuring the long-term success of the project, although the 2% stakes in WAPCo of the Société Beninoise de Gaz and the Société Togolaise de Gaz could (again, purely for the sake of argument) be regarded as insignificant if a serious need for off-take renegotiations (or the introduction of a transit fee) arose as far as either transit state was concerned.

External influence from international institutions

Such factors as IMF support for economic development to the Georgian and Turkish governments and the significance (or otherwise) Turkey assigns to accession to the EU could play a key role in the success of the BTC project, in terms of the potential of these countries to enforce their bargaining advantage in contract renegotiations. In the case of the WAGP and Chad–Cameroon projects, the World Bank was significantly involved in various economic development projects in the region, especially with the transit countries.

The involvement of international finance organisations in the economies of the parties to these pipeline projects (in addition to these countries' affiliations to numerous inter-governmental organisations) fosters *mutual dependency* between the parties. Chapter 6 further pursues mutual dependency as a workable concept for solving the problem of potential interruptions owing to renegotiation of transit agreements.

4.4.2   The patience principle and the risk-aversion principle: bargaining positions

In Chapter 3 it was established that there are time implications for the parties to a pipeline agreement, such that the party which incurs less cost by prolonging negotiations or renegotiations is the one with more bargaining power. It was also argued that the bargaining power of transit countries would be dependent on international relations between the countries involved as well as the contribution of the parties to the cost of the project. The value of the four projects to the producer parties is obvious, as pointed out in preceding sections of this chapter. The question, therefore, is whether the transit countries can afford to prolong renegotiations for improved terms upon commencement of pipeline operations. This question can be said to have been answered already in section 4.3 on the value of the projects to these countries.

In the BTC and Chad–Cameroon oil pipeline cases, Georgia, Turkey, and Cameroon stand to benefit significantly from the pipelines in terms of revenue and FDI. In the WAGP and Shah Deniz/SCP gas pipeline cases, the transit countries have energy requirements for the off-take gas.

It can, therefore, be argued that restricting supply through the pipelines would have as much negative impact on the transit countries as on the producer states, especially in the case of gas pipelines.

With the gas projects, the likelihood of arbitrary contract renegotiations by Georgia, Turkey, Benin, or Togo can be argued to be low, given the potential cost to these countries' power and industrial projects of interrupting the operation of the pipelines. In the case of the oil pipelines, however, the transit countries cannot be expected not to arbitrarily seek renegotiations simply because of the potential benefit in terms of oil rent. For the purposes of this research they can be argued to be pure transit countries, and it can, therefore, be said, *ceteris paribus*, that on the basis of potential economic rent alone, they could seek further renegotiations when the pipelines are operational.

International relations in these cases can be argued to have little impact on the bargaining positions of the transit countries. In the case of the WAGP project, for example, although the internal economic and political positions of Benin and Togo can be said to be unstable, Nigeria has significant economic and political influence over the West African region, and these countries have economic dependencies on Nigeria outside of the pipeline. The implication is simple: international disputes between these countries are unlikely to trigger interruptions to the pipeline. In the case of the BTC and SCP projects, the Georgia–Turkey route was chosen over Russia, Armenia, and Iran specifically to avoid such disputes, although the Russia option may have been excluded to reduce its monopoly over access to Caspian energy. The parties to these pipeline projects can, therefore, be said to be (as far as can be measured) equally risk averse, thus reducing the likelihood of interruptions to the pipeline as a result of transit disputes. Their broader objectives can be said to be similar, given the roles they play in the energy corridor.

Getting transit countries to contribute to the cost of a pipeline project can also reduce the likelihood of arbitrary renegotiation of transit terms, although this would in turn increase their exposure to project risks. There is some cost contribution by each of the transit countries in the case studies. Just as the stakeholders of any company are committed to ensuring its success, the transit parties, by way of cost contribution, are expected to ensure the security of supply of oil or gas through the pipelines.

In conclusion, the transit countries in these cases cannot be said to be less risk averse than the producer countries. It can also be concluded that there are significant cost implications of arbitrary renegotiation to these countries.

4.4.3 The inside and outside options principle: pipeline alternatives

The application of the inside and outside options principle to the case studies is straightforward. In the case of the gas pipelines, cost plays a significant role in determining the outside options available to the parties. There is simply no feasible alternative to the WAGP to transport gas to Ghana, thus shifting bargaining power to Benin and Togo. However, inside options for power generation by gas in these transit countries are more expensive than the off-take option. The implication for their positions in terms of an arbitrary renegotiation for more off-take (or for a transit fee in addition to the off-take) is one of arguable stability; that is, the project is too important for them to interrupt.

With the Shah Deniz/SCP project, the choice of the pipeline route was strongly influenced by the market for the gas. Turkey was identified as having the market capacity, economic growth, and financial ability to take and pay for the gas. Turkey already owns and operates a pipeline, which has positive cost implications for the project because the SCP will connect to this pipeline at the Turkish border, reducing construction costs. Again for this case, there is no immediate feasible outside option available for the export of Azeri gas to Western markets. However, the off-take benefits to Georgia and the countries' stated willingness to support the project can be argued to even out the bargaining positions of the parties to the project.

In the case of the BTC pipeline, there is an existing network of pipelines to Western markets in the Caspian region through which Azeri crude can be exported. Bargaining power could, therefore, be argued to favour the producer country, since there are alternative routes to the BTC pipeline. However, there would be capacity issues with the other pipelines should the need arise to pipe BTC oil through them, which affects the commercial feasibility of these alternative options. In addition, there is strong Western (particularly US) influence over the project. The need to ensure a secure supply of Caspian oil to Western markets (other than via the Russian route) has influenced US-encouraged projects such as the BTC pipeline. Georgia is seen as a vital component of the East–West Energy Corridor. The implication is that the BTC pipeline can compete favourably with the alternative options available for the transportation of Azeri crude to Western markets. Moreover, the alternative options available to both Georgia and Turkey do not compare favourably with the BTC pipeline.

The case of the Chad–Cameroon pipeline is also straightforward. There is no commercially feasible alternative. The pipeline, however, also benefits Cameroon, both through rent-sharing and in terms of the development prospects for the oil fields of the Logoni Basin (an area that would otherwise be expensive to develop) simply by connecting to the Chad–Cameroon pipeline.

4.4.4 The commitment tactics principle: the cost of revoking agreements

From the review of the commitment tactics principle in Chapter 3, the key question is what impact the cost of revoking commitments would have on the bargaining position of the parties involved. The review suggests that the party for which the cost of revoking its partial commitment is higher has more bargaining power. It is, therefore, important to examine the relevant legal obligations of the parties to the project, with a view to investigating what the costs would be to the transit country of revoking such commitments.

The parties to the BTC project are legally bound by an inter-governmental agreement and host government agreements signed by the three participating countries (Azerbaijan, Georgia, and Turkey). The agreement gives legal backing and recognition to the project and its participants, and stipulates their roles and obligations. It is ratified by each party's parliament and deemed effective under the country's constitution and legal regime. The agreement obliges all party states to fulfil their duties and obligations arising under any project agreement. Permission is also granted for all necessary measures to be taken to avoid delays and operational difficulties that would affect the project, from construction through to operation and maintenance. In accordance with the objectives of the ECT, section 4(viii) of Article II of the agreement stipulates free movement of goods, materials, supplies, technology, and personnel to and among the facilities, free of all taxes and restrictions. Section 9 of Article II states (importantly) that the agreement does not affect the right of each state to enter into any other project agreements or to modify or terminate any project agreement, apart from this agreement.

The general thrust of such agreements (and it must be pointed out that more substantive detail, such as the transit fees paid or the exact off-take volumes, is sometimes excluded for reasons of commercial sensitivity) is to document the legal obligations of the parties to the project. Chapters 2 and 3 establish that, *ceteris paribus*, the transit country might seek to renegotiate (in other words, breach the terms of the pipeline agreement) simply because it can (the obsolescing bargain). However, such renegotiation will be subject to the cost of the transit country's revoking other agreements related or unrelated to the pipeline project. Turkey's EU accession aspirations are an example, with respect to the BTC and SCP projects; so is Georgia's relationship with the US as an ally. In the case of the WAGP, the cost to each transit country of revoking commitments in terms of its ability to meet domestic energy requirements would have to be considered before deciding to renegotiate the pipeline agreement for either higher off-take or a transit fee in addition to the gas off-take.

4.5 Conclusion

In terms of underlying motives, the construction and operation of the BTC pipeline has potential benefits for US and regional policy objectives such as security of supply, unlocking the Caspian deposits, and ensuring that neither Russia nor Iran has a monopoly over pipelines in the region. It is a competitive alternative option to the NREP and WREP projects. The same can be argued for the motives behind the SCP project. One interesting aspect of the four cases is the significant international presence, in terms of financial support from the World Bank, the IMF, and other international financial institutions.

This chapter has shown that the more relevant the project is to all participants, the more interest there is in its success. Equity participation in the construction, operation, or maintenance of the project could reduce the likelihood of arbitrary renegotiation. In terms of applying the principles of bargaining theory to the case studies, this chapter has found that it is important for the transit countries not to have more bargaining power than the other parties (the producer countries) if disruptions to the pipeline project are to be avoided. Pure bargaining may not be sufficient to prevent arbitrary renegotiation once the pipeline is operational. In fact, the essence of naked bargaining is to enhance the position of the party with more bargaining power, which in these cases would significantly affect the pipeline project. A situation such as the Chad–Cameroon pipeline case, where the transit fee does not reflect fluctuations in oil price, only strengthens the tendency for the transit country to squeeze more rent from the project. Leaving the determination of the transit fee and the renegotiations that could follow to the relative bargaining positions of the parties will not yield a reasonable, objective, and non-discriminatory transit pipeline agreement. Instead, what it does is encourage the transit country constantly to squeeze rent from the pipeline project, until the project is significantly disrupted. This cannot be said to be a reasonable, objective, and non-discriminatory process. The impact of relative bargaining power in transit pipeline negotiations is very powerful. What the cases analysed in this chapter do show, however, is that certain actions, factors, or even circumstances can serve to *counter-balance* this bargaining power such that there is less likelihood of arbitrary rent-squeezing. This, of course, also cannot be said to be reasonable, non-discriminatory, or fair (if one dare be so vague), because, drawing again on the Chad–Cameroon example, it is not impossible that Cameroon will find the transit terms unacceptable a few years into the operation of the pipeline, despite the potential benefits the available evidence suggests it will accrue from the pipeline.

Points raised in this chapter highlight possible ways in which the relative bargaining power of the parties to the agreement can be maintained so that arbitrary use of naked bargaining power does not occur. A common interest to participate in the pipeline works favourably towards such an objective. In terms of the patience principle, it is important that the costs of disruptions to the pipeline are significant to the transit and producer countries alike (if not the same for both). If there are competitive alternatives to the transit route, the tendency of the transit country to seek renegotiations is reduced. Finally, given the commitment tactics principle, the benefits of tying the transit country to other partial commitments directly or indirectly related to the pipeline are obvious. The US involvement in the BTC project, the role of Georgia in the East–West Energy Corridor, and the potential benefits if the WAGP continues across the ECOWAS states are examples of ways in which the bargaining positions of transit countries can be managed. These points serve to shape the direction of Chapter 6, which addresses the concept of mutual dependency between the parties to the pipeline as a mitigation for the deficiencies of naked bargaining theory in pipeline agreements.

Note

1. It is important to state that these indices are based on 2008 rankings only, and they tend to vary from year to year. However, movements up or down the rankings (especially for the countries with high indices) are not dramatic.

5
The Role of the Energy Charter Treaty: A Critique

Abstract: *The previous chapter showed that the motives behind the construction and operation of a pipeline, the relevance of the project to the participants, equity participation in the project, and the availability of alternatives are all important factors in the balance of bargaining power among the parties involved. The problem highlighted at the beginning of this research is that the Energy Charter Treaty and the draft Energy Charter Protocol on Energy Transit cannot overcome the possibility of opportunism. The extent to which the factors identified in Chapter 4 are incorporated into the philosophy of the Treaty as a whole is not clear at this point. The objective of this chapter, therefore, is to analyse the relevant provisions of the Treaty and the draft Protocol in the context of the findings from this research, with a view to characterising the role of the Energy Charter Secretariat (or indeed any form of international regulation) in cross-border oil and gas pipelines involving transit.*

Omonbude, Ekpen James. *Cross-border Oil and Gas Pipelines and the Role of the Transit Country: Economics, Challenges, and Solutions.* Basingstoke: Palgrave Macmillan, 2013. DOI: 10.1057/9781137274526.

5.1 Introduction

The previous chapters argued that the position of the transit country after a pipeline has been built and put into operation is significant. The transit country could seek to optimise rent from the pipeline simply because it is able to do so. An analysis of bargaining principles in the context of oil and gas pipelines and the four pipeline case studies revealed factors that influence the positions of the participants and factors that could serve to check the swings in bargaining power, especially in favour of the transit country. The previous chapter showed that the motives behind the construction and operation of a pipeline, the relevance of the project to the participants, equity participation in the project, and the availability of alternatives are all important factors in the balance of bargaining power among the parties involved.

One major problem regarding cross-border oil and gas pipelines – as highlighted in Chapter 1 – is the lack of a regulatory regime that adequately addresses transit problems with cross-border oil and gas trade. With the exception of World Trade Organization (WTO)/GATT law,[1] the ECT is the main regulatory and dispute settlement instrument for the transit of energy commodities. General provisions in Article 7 of the Treaty are designed to address the transit of oil and gas, and the draft ECTP is intended to address the issues in greater detail.

The problem highlighted at the beginning of this research is that the Treaty and the draft Protocol cannot overcome the possibility of opportunism. The extent to which the findings in Chapter 4 concerning the factors that have a significant impact on the bargaining positions of the parties to the pipeline are incorporated into the philosophy of the Treaty as a whole is not clear at this point. The objective of this chapter, therefore, is to analyse the relevant provisions of the Treaty and the draft Protocol in the context of the findings from this research, with a view to characterising the role of the Energy Charter Secretariat (or indeed any form of international regulation) in cross-border oil and gas pipelines involving transit.

This chapter looks at the background of the ECT and draft Protocol on transit, as well as their relevant provisions. It then analyses the role of these international regulatory instruments in the context of the consequences of shifts in bargaining power.

5.2 The origin and objectives of the ECT and the ECTP

5.2.1 The ECT

The ECT is described as the first economic agreement that attempts to unite the former Soviet Union, the formerly centrally planned economies of Central and Eastern Europe, and the Organization for Economic Cooperation and Development, with the exception of the US, Canada, Mexico, and New Zealand (Energy Charter Secretariat, 1996). It is the first binding multilateral investment protection agreement, in addition to being the first agreement that covers both investment protection and trade and, more importantly, the application of transit rules to energy networks. The ECT was signed in Lisbon in 1994, after four years of what can be termed the first phase of the Treaty's development. The European Energy Charter was originally conceived as a means of advancing the complementary relationship in terms of energy matters between the former Soviet Union, Central and Eastern Europe, and the West (Waelde, 1996). The European Energy Charter was a declaration that was not considered to be a legally binding international treaty. The 1994 ECT gave it legally binding status (Bamberger et al., 2000).

The major reason behind the negotiations leading to the creation of the European Energy Charter and eventually the ECT can be argued to be the effects of the collapse of the former Soviet Union in 1990, after which the call was made for a European energy charter that would constitute a political and legal foundation for economic cooperation between the East and the West. This cooperation, it was argued, would support the general transition to a market economy and enhance European political stability through the promotion of Eastern economic development. In addition, it was suggested that Western energy dependence and security of supply issues would be addressed by the creation of an advantaged relationship of investment and trade between East and West. The broad assumption was that if Eastern European economies provided or promoted investment security, the result would be higher investment inflows, which in turn would develop these Eastern economies and supply energy to the West. The proposed liberation of the energy market from regulatory hindrances – it was assumed – would enhance the movement of Western finance and technology eastward, while Eastern products (especially energy) would move westward (Bamberger et al., 2000).

It has been argued that substantial parts of the ECT borrowed from three major sources (Waelde, 2004):

- existing bilateral and multilateral investment agreements (e.g. the North American Free Trade Agreement and the aborted negotiations for a Multilateral Agreement on Investment);
- GATT, especially on trade and transit issues; and
- what Waelde refers to as the 'liberalisation fad': motivation from several directives and new regulatory regimes concerning such issues as non-discriminatory access to energy transport infrastructure, transit, and a number of energy-related issues.

In terms of institutional acceptance, the notion of such East–West cooperation was welcomed by the European Council, and thus a study commenced on how best to implement it. In February 1991 the Commission proposed the concept of the European Energy Charter. This was followed by the European Union's invitation to the other countries of Western and Eastern Europe, the former Soviet Union, and the non-European members of the Organization for Economic Cooperation and Development to attend a conference in Brussels in July 1991, the principal aim of which was to begin negotiations on the European Energy Charter.

The first non-European states to join these negotiations were the US, Canada, Australia, and then Japan. What emerged from these 'first-generation' negotiations was a statement of major common principles that signatory states wished to pursue, in addition to a set of procedures or pointers for the subsequent negotiation of a 'Basic Agreement' (which would later be referred to as the 'Energy Charter Treaty') and a set of actions (or 'Protocols', as they are referred to in the Charter).

Key milestones in the history of the ECT include its coming into effect in 1998 with 30 signatories, as well as the Protocol on Energy Efficiency. (Figure 5.1 shows the signatories as of January 2003.) Negotiations commenced in 2000 on the Transit Protocol and were originally expected to conclude in December 2003. However, the negotiations were suspended, resumed in June 2004, and again experienced difficulties, especially with securing Russia's ratification.

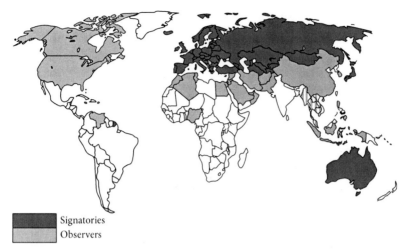

FIGURE 5.1 *Signatories to the Energy Charter Treaty (as of January 2003)*
Source: Energy Charter Secretariat website.

5.2.2 The ECTP

According to the Energy Charter Secretariat, the ECT's existing transit provisions oblige signatory states, in accordance with the GATT/WTO principles of freedom of transit, to facilitate transit on a non-discriminatory basis. However, after the adoption of the Treaty in 1994, increasing consensus emerged within the Energy Charter Conference that the provisions on energy transit needed to be enhanced and strengthened. More specifically, it became increasingly obvious that the collapse of the Soviet Union had created specific energy transportation problems that could significantly impede the development of energy projects involving cross-border transit in the CIS region. This has led to frequent problems over energy transit flows in a number of former Soviet Union states over recent years (such as the much-discussed Russian–Ukrainian gas disputes; see also Dodsworth *et al.*, 2002; Stern, 2006, 2009).

It was, therefore, agreed that the members of the Energy Charter Conference should aim to establish a clearer set of rules for inter-state transit flows. As a result, in December 1999, the Conference authorised the commencement of negotiations on the ECTP, which was intended to strengthen the existing transit-related obligations on governments under the ECT.

The ECTP is being developed according to the importance attached (by countries in Europe and Asia especially) to establishing clear rules governing energy transit across boundaries. The primary objective of the Protocol is to create a legally binding multilateral agreement that sets out rules for cross-border flows of energy in transit (Kemper, 2003a). It aims at the development of a regime of commonly accepted (legal) principles covering transit flows of energy resources, both hydrocarbons and electricity, crossing at least two national boundaries. The Energy Charter Secretariat argues that, in creating these rules (and abiding by them), signatory states are likely to benefit from increased political and legal stability for energy transit and reduced risks in oil and gas projects. The provisions in Article 7 of the Treaty regarding energy transit (discussed below) are quite general and require further elaboration to reduce disputes arising from different interpretations. The Protocol is intended to specify these provisions more clearly.

5.3 Significant provisions of the ECT

The ECT provisions come under four main categories: (1) investment protection and promotion, (2) energy trade, (3) energy efficiency, and (4) transit. This section summarises the general provisions of the Treaty under these headings and then proceeds to discuss the provisions of the ECTP. The purpose of discussing the general provisions of the Treaty (instead of discussing the transit provisions alone) is to highlight the comparative ease with which agreements regarding these provisions have been reached.

5.3.1 Investment protection and promotion

The provisions for investment protection and promotion predominantly concern investments from outside parties associated with an 'economic activity in the energy sector'. In terms of scope, they cover the entire production process: exploration, extraction, refining, production, storage, land transport, transmission, distribution, and trade of energy materials and products.

In addition to rules on employment requirements, compensation for losses and asset expropriation, and currency transfer, the investment protection and promotion provisions of the Treaty establish (similar to most bilateral investment agreements) the notion of 'national treatment'; that is, foreign investors should be treated at least as well as domestic investors.

The economic importance of national treatment is that it increases the propensity of investors to invest in countries covered by the Treaty, since – in the ideal environment determined by the Treaty's investment provisions – they will not be discriminated against.

5.3.2   Energy trade

The ECT relies generally on the GATT 1947 rules and related instruments for trade. This bears great significance with respect to the Treaty's outreach. By basing its provisions on GATT 1947, the Treaty incorporates transition economies that are not WTO members but are determined to participate in the international institutions of the world market economy.

5.3.3   The Protocol on Energy Efficiency

Energy efficiency as a concept started to become a factor in energy policies following the first oil shock of the early 1970s. The emphasis during this period was on security of supply and the balancing of supply with demand for oil. The concept of energy efficiency has since broadened to include transition to a market economy and environmental and climate change issues. The Energy Charter Protocol on Energy Efficiency and Related Environmental Aspects was negotiated in 1998 and came into force the same year. The Protocol serves as the legally binding instrument in the area of energy efficiency and establishes governmental requirements for the formulation and implementation of policies and programmes.

The primary objectives of the Protocol are as follows (Energy Charter Secretariat, 1996):

- to promote energy efficiency using policies consistent with sustainable development;
- to create a framework of conditions which encourage economical, efficient, and environmentally sound energy use by producers and consumers alike; and
- to promote cooperation in the area of energy efficiency.

The provisions of the Protocol include policies that promote the efficient functioning of market mechanisms (such the reduction of barriers to energy efficiency), access to private capital markets and international financial institutions, and energy-efficient technology.

The dispute settlement procedures in the Treaty regarding the provisions on investment, trade, and energy efficiency are straightforward.

For example, there are clear provisions for courts and administrative tribunals for the settlement of disputes arising from a breach, and, by virtue of being a signatory to the Treaty, any contracting party gives unconditional consent to the submission of disputes relating to the above provisions to international arbitration or conciliation in accordance with the Treaty's provisions. There are also other clearly defined international legal instruments for the settlement of such disputes (e.g. the International Centre for Settlement of Investment Disputes Convention) which have been used successfully and can be applied to the energy-specific cases arising from breaching the provisions discussed in sections 5.3.1–5.3.3.

As is shown from section 5.3.4 onwards, such clarity is lacking for the Treaty's transit provisions. This is because the consequences of disputes arising from transit are quite different from the consequences of a breach of the provisions on investment, trade, or energy efficiency, in terms of the effects such disputes have on time lags, costs, and the price and, ultimately, the security of supply of the commodity.

5.3.4   Transit

Of the four major classifications of the ECT provisions, transit is the most relevant to this research. It is also from the general provisions in this category that the ECTP has been developed. Transit is defined in the Treaty as the movement of energy commodities from one contracting party, through the territory of another contracting party, to a third country (Waern, 2002). Article 7 of the ECT outlines the provisions regarding the transit of energy products and materials to and from signatory and observer states to the Treaty.

Article 7(10) of the ECT defines transit in two ways: first, essentially as the carriage of energy materials and products from one area to another through a 'transit' area or state; second, as the carriage of these energy materials simply from one area to another, unless these two areas (Contracting Parties) decide otherwise. (See Box 5.1.)

Freedom of transit

The principle of freedom of transit is upheld in Article 7 of the Treaty, in which each contracting party is expected to take the necessary measures to facilitate the transportation of energy materials without distinction as to origin, destination, and ownership of these commodities. In terms of the

BOX 5.1 *Definitions of transit in Article 7(10)(a) of the Energy Charter Treaty*

(a) Transit means:
 (i) the carriage through the Area of a Contracting Party, or to or from port facilities in its Area for loading or unloading, of Energy Materials and Products originating in the Area of another state and destined for the Area of a third state, so long as either the other state or the third state is a Contracting Party; or
 (ii) the carriage through the Area of a Contracting Party of Energy Materials and Products originating in the Area of another Contracting Party and destined for the Area of that other Contracting Party, unless the two Contracting Parties concerned decide otherwise and record their decision by a joint entry in Annex N.

Source: Energy Charter Secretariat (1996).

pricing of these energy materials, the Treaty expects non-discrimination on the same basis: no distinctions as to origin, destination, and ownership.

The aim here is to achieve one of the primary objectives behind the creation of the Treaty in the first place: relatively unhampered access to energy resources. The desired effects of this access range from security of supply to the argued benefits of a market-determined energy sector. The Treaty makes provision for the protection of energy commodities in transit in the event of disputes. It expects the contracting party through whose area energy materials and products transit not to interrupt or reduce the existing flow of energy materials and products in the event of a dispute over any transit issues.

Energy transport facilities

Paragraphs 2 and 5 of Article 7 promote the modernisation of energy transport facilities essential to energy transit. Paragraph 2 states that cooperation is expected from contracting parties regarding the development and operation of such transport facilities, and that contracting parties are expected to mitigate the effects of supply interruptions and facilitate the interconnection of these facilities. The simple implication of this is enhanced transit of energy materials. It is important to note, however, that the transit country is not obliged to permit the construction or modification of these facilities. The transit country is also not obliged to permit additional transit through existing facilities. Decisions not to allow further transit or facility modifications are usually attributed to such risks as endangering the security or efficiency of existing transport facilities (Waern, 2002).

Fair treatment

Under paragraphs 3 and 4 of Article 7, contracting parties are expected to make provisions relating to the transport of energy such that energy in transit is treated 'no less favourably' than either energy produced domestically for export or imported energy. Paragraph 4 expects contracting parties not to place undue or unfair obstacles in the way of new capacity being established in the event of existing transport facilities not being able to cope with commercial pressures. This calls on such concepts as 'non-discrimination', 'unreasonable restrictions', and 'objectiveness'.

These are relatively well-known concepts in the sphere of international investments and the legal context surrounding them, which allows common legal interpretation of the concepts. What is of interest, however, is the economic and philosophical interpretation of the constituents and characteristics of these concepts for transit issues in energy economics (i.e. the how, why, what, who, and where).

5.4 Significant provisions of the draft ECTP

As already mentioned, the provisions of the draft Protocol are intended to be more specific than the provisions in Article 7 of the ECT. According to the Energy Charter Secretariat, the main objectives of the Protocol are:

- secure transit for all parties' benefit;
- promotion of transparent and non-discriminatory access to capacity;
- efficient use of transport facilities; and
- prompt and effective dispute settlement.

This section outlines the significant provisions with respect to the scope and jurisdiction of the Protocol, the nature of agreements, the taking of energy in transit, capacity utilisation, transit tariffs, and other charges.

5.4.1 Provisions on the scope of the Protocol and transit agreements

Articles 4(3) and 5 cover the scope of the draft Protocol and transit agreements. Article 4(3) reaffirms the rights of the contracting parties under international law. It states that the rights of any contracting party to receive fair and reasonable benefit for its role in the construction, expansion, extension, and reconstruction of energy transport facilities used for transit within its territory are not restricted by the Protocol.

Article 5 expects that each contracting party will observe its obligations resulting from transit agreements with other entities of contracting parties. What is of note here is the provision that no transit agreement concluded before the ECTP comes into force shall be challenged as being in violation of any provision of the Protocol. This could have problematic implications for the Protocol's objective of quick resolution of potential disputes relating to pipeline agreements concluded prior to the enforcement of the ECTP.

5.4.2 Provisions on energy in transit and available capacity

Under Article 6, the transit country is expected not to engage in unauthorised off-take of energy, and it is the responsibility of the transit country to ensure the prohibition of illegal off-taking. With respect to the utilisation of available capacity, Article 8(1) covers the negotiation, in good faith, of owners or operators of energy transport facilities with other contracting parties. However, the transit country has the right to cease or prohibit negotiation if it does not maintain a diplomatic relationship with the other contracting parties; it can also prohibit transactions with entities of a state with which it does not have positive diplomatic relations.

5.4.3 Provisions on transit tariffs and other charges

The provisions on tariffs are in Article 10 of the draft Protocol. Transit tariffs and other conditions, according to this article, must be objective, reasonable, transparent, and non-discriminatory with respect to the origin, destination, and ownership of energy materials. The article expects that tariffs and other conditions will not be affected by market distortions, 'in particular those resulting from abuse of a dominant position by any owner or operator of energy transport facilities used for transit'. Tariffs are expected by the Protocol to be designed on the basis of operational and investment costs, in addition to a 'reasonable' rate-of-return framework.

The draft Protocol does not explicitly state whether 'other charges' refers to transit fees. However, since there is a difference between the transit tariff charged by the pipeline operator and the transit fee paid to the transit country, it can be assumed that the draft Protocol is referring to the transit fee in its provision for other charges. It specifically stipulates in Article 11 that 'any charges imposed by a contracting party on transit of energy materials and products within its territory shall fully comply with Article V of GATT 1994'. GATT Article V makes provisions for such issues as the freedom of transit. Transit is expected to be via the routes most convenient for international transit, without distinction as to place of origin or departure, entry or exit, or ownership. In addition, all charges in any transit agreement are expected to be reasonable, with the additional provision that no party or commodity will be treated less favourably than another in transit.

5.5   Analysis of the role of the ECT and ECTP

As stated in section 5.1, the purpose of this chapter is to analyse the role of the instruments of the Energy Charter Secretariat in alleviating the problems considered in this book. In terms of international regulation, the ECT and the draft ECTP are the instruments most relevant to transit issues in cross-border oil and gas pipelines. It is, therefore, pertinent to analyse their role in the resolution of the problems that arise from the transportation of these commodities via pipeline and, more specifically, the extent to which they address the consequences of shifts in the relative bargaining power of the parties to transit agreements. This entails assessing whether the provisions of the ECT or draft ECTP address the influential factors in the shift of bargaining power discussed in Chapter 3 and applied to the case studies in Chapter 4.

It was established in Chapters 3 and 4 that the potential for arbitrary interruption of the pipeline by the transit country for rent-squeezing is essentially a consequence of shifts in bargaining power to the transit country. It can be argued that this is a solution to the bargaining problem, since leaving it to pure bargaining yields an outcome, albeit one that is favourable to the party with greater bargaining power. However, taking security of supply into consideration, it is important that such shifts in power, which can encourage aggressive behaviour by the transit country, are muted. The provisions of the ECT and ECTP do not sufficiently address this problem:

- Article 7(1) of the ECT and Article 10 of the draft Protocol uphold the principle of freedom of transit and dictate non-discrimination on the basis of origin, destination, and ownership of commodities and transport infrastructure. This, however, does not guarantee absolute freedom of transit of the commodities, simply because disputes can still arise over the transit fees, contractual obligations, and so on that could obstruct the transit of energy from origin to destination.
- Under Article 7(2) and (5) of the ECT, the transit country is not obliged to allow the construction or modification of energy transport facilities. Article 8(1) of the draft Protocol recognises the right of the transit country to cease or prohibit negotiations with countries with which it has unfavourable diplomatic relations. These provisions protect the rights of the transit country, and thus enhance its bargaining power. They give the transit country the upper hand – should the need arise for additional transit volumes or modification of transport facilities – in terms of transit fee renegotiations, for example. Provided all the contracting parties agree to be bound by the provisions of Article 7(5) of the ECT, the transit country apparently stands in a better position, since additional transit could be conditioned upon meeting new demands raised by the transit country. This, in turn, could affect the security of supply of the energy commodity in question, as disputes can arise from such new demands. Although Article 7(6) of the ECT can be argued to address this problem, as the transit country is expected to continue to allow existing transit even during such disputes, there are no clearly stated enforcement mechanisms to ensure such uninterrupted transit.
- Paragraphs 3 and 4 of ECT Article 7 on the notion of 'fair treatment', as well as Article 11 of the draft Protocol, are vague. The use of the terms 'reasonable', 'transparent', 'objective', and 'non-discriminatory' to describe transit tariffs and other charges (which will include the transit fee) cannot wholly apply to cross-border oil and gas pipelines, as Chapters 2–4 demonstrate. Again, this is simply due to the bargaining problem. A transit country's justification for exercising its bargaining power in renegotiating transit terms can be deemed unreasonable by the other parties, and vice versa. The definition of reasonableness, objectiveness, and so on is relative, and

in the case of pipelines involving transit it depends strongly on each party's motive for participating in the project.
- Nothing in the provisions of the Treaty or the draft Protocol pronounces strict punitive measures that could deter aggressive behaviour by any of the parties to the pipeline agreement, especially rent-squeezing by the transit country.
- Article 5 of the draft Protocol states that no agreement concluded prior to the coming into force of the ECTP can be challenged as being in violation of any of the ECTP's provisions. This raises the question of what the implication is for disputes arising in virtually every current transit pipeline agreement within the Protocol's jurisdiction (as the Protocol is yet to come into force), and how much jurisdiction the ECTP will have over such pipelines.
- The Charter is still unable to obtain an agreement between Russia and the EU on a number of issues, specifically the provisions which regulate third-party access to oil and gas pipelines (under such provisions as the regional integration clause, right of first refusal, and transit tariffs).
- The draft Protocol suggests in Article 10(3) the calculation of tariffs and other charges on the basis of operational and investment costs and a reasonable rate of return. This clearly shows the lack of a full appreciation of the role of the pipeline transit country. So long as the full value (or cost) of the pipeline to the transit country is not captured by the above methods (and other methodologies such as distance-based and entry–exit tariffs), such calculations cannot be said to bear the characteristics of fairness in the ECT context.

The viewpoint of the Energy Charter Secretariat is also that the legal instruments (i.e. the ECT and draft Protocol) are struggling to address the potential for disputes to arise from disruptions to a pipeline as a result of the transit country renegotiating the agreement. From a number of discussions with a senior official of the Secretariat, it seems that there is an acknowledgement of the risk of transit countries adopting a rent blackmailing position. It is anticipated that the principle of freedom of transit (as transcribed from GATT Article V), which expects transit countries not to discriminate or to charge fees that are not justified by costs, will hold. The draft Protocol expects transit countries not to create obstacles to transit and states that they can raise the same tax income as from any similar business. The official also argued that the state-to-state dispute settlement procedures under Article 27 of the ECT and Article 21 of the draft Protocol (which provide for an *ad hoc* tribunal under United Nations Commission on International Trade Law rules) should address the problem; however, negotiations for a new transit route can be very complex, and there are no specific mechanisms by which threats of arbitrary rent-squeezing by the transit country can be prevented or eliminated.

5.6 Conclusion

The review of the provisions of the ECT and draft ECTP in sections 5.3 and 5.4 and the analysis of the roles of these instruments in section 5.5 clearly show that the Treaty and draft Protocol do not adequately address the potential problem of disputes arising from the arbitrary interruption of an operating pipeline by the transit country. It is evidently not sufficient to have binding obligations on signatory states to prevent illegal taking of energy in transit, as there are no clear punitive measures in place for defaulting. It is also not sufficient to state that nothing affects the right of the transit country to receive benefits for facilitating transit (although restricting such rights would impinge on the transit country's sovereignty) and not to set limits on what it is reasonable for the transit country to collect (although such limits are difficult to determine and could also be deemed *unreasonable*). Moreover, given that the Protocol has not yet entered into force, there is little or nothing it could do to settle disputes regarding transit fees that arose from the pipeline agreements which it excludes from its coverage. This is not helped by the fact that none of the pipeline agreements (within the Protocol's jurisdiction) made prior to the entry into law of the Protocol can be challenged as being in violation of its provisions. In addition, there is no clear definition of non-discriminatory access to available transit capacity as far as the Treaty or the draft Protocol is concerned.

In his recommendation for the way forward, a senior official of the Energy Charter Secretariat suggested that an ideal solution would be one that combined 'legal instruments' and '*de facto* instruments' – the legal instruments being the ECT, the ECTP, and aspects of WTO and GATT law, and the *de facto* instruments being international institutions such as the World Bank (via conditions attached to loans and other World Bank instruments). This implies that in addition to the factors that affect the outcome of the bargaining process, there are factors outside the pipeline agreement that can serve to check the consequences of shifts in bargaining power (to prevent arbitrary renegotiation of transit terms by the transit country). These are analysed in the following chapter.

Note

1. The role of WTO/GATT law is not specifically analysed in this book, because the provisions of the ECT and the draft ECTP are essentially derived from WTO/GATT provisions, making a separate analysis of the WTO/GATT provisions unnecessary.

6
A Case for Mutual Dependencies

Abstract: *The transit country will always seek to expropriate the total rent arising from the pipeline. Shifts in bargaining power among the parties to the pipeline agreement affect the transit country's ability to influence transit fees, depending on the factors that prompt these shifts. Bargaining, nevertheless, can be argued to be the only platform upon which a solution to the transit fee dispute puzzle can be developed. The essential requirement, therefore, is an institutional framework that obviates expropriation. This chapter investigates the aspects of mutual dependencies among the parties to the pipeline agreement that will mitigate the consequences seen when the bargaining principles are applied to the case studies. It identifies the context in which mutual dependency among the parties to the pipeline project is demonstrated with regard to transit country bargaining power (i.e. factors in the transit agreement that demonstrate such dependencies and factors outside the agreement that could also affect the bargaining power of the transit country).*

Omonbude, Ekpen James. *Cross-border Oil and Gas Pipelines and the Role of the Transit Country: Economics, Challenges, and Solutions.* Basingstoke: Palgrave Macmillan, 2013. DOI: 10.1057/9781137274526.

6.1 Introduction

Chapters 2 and 3 showed that for cross-border oil and gas pipelines passing through transit countries, it is impossible to have a transit agreement that bears the characteristics of reasonableness, objectiveness, transparency, and non-discrimination (in the long term), as suggested by the ECT, which was analysed in Chapter 5. The transit country will always seek to expropriate the total rent arising from the pipeline. Shifts in bargaining power among the parties to the pipeline agreement affect the transit country's ability to influence transit fees, depending on the factors that prompt these shifts. The principles of bargaining were applied to the four pipeline cases in Chapter 4, and the analysis underscored the difficulty of achieving a long-term pipeline agreement that is wholly acceptable to all the parties involved. The analysis also showed how the bargaining position of the transit country is affected by factors surrounding the entire project, and how this position can change over time as a result of exogenous factors directly or indirectly related to the project.

This analysis can be argued to be straightforward, given the way in which such factors as the motive behind the project, the importance of the project to the transit country, and geo-politics seem to fit directly into the bargaining principles framework set out in Chapter 3.

However, a problem arises with this straightforward application. It is theoretically difficult (if not impossible) to have a pipeline agreement such that chances of arbitrary renegotiation of transit terms by the transit country are reduced, because any agreement will be subject to the sovereign powers of the parties involved. Therefore, it is not sufficient to suggest pure bargaining as a stand-alone solution. From a pure bargaining perspective, the bargaining outcome can be deemed *fair*; can be deemed reasonable and objective (as far as there is acceptance of the limitations inherent in such pure bargaining forces). From a security of supply perspective, however, the impact of disputes from arbitrary rent-squeezing is huge, both for the exporter country and for the destination market.

Bargaining, nevertheless, can be argued to be the only platform upon which a solution to the transit fee dispute puzzle can be developed. The essential requirement, therefore, is an institutional framework that obviates expropriation. This chapter investigates the aspects of mutual dependencies among the parties to the pipeline agreement that will mitigate the consequences seen when the bargaining principles are applied to the case studies. It identifies the context in which mutual dependency among the parties to the pipeline project is demonstrated with regard to transit country bargaining power (i.e. factors in the transit agreement that demonstrate such dependencies and factors outside the agreement that could also affect the bargaining power of the transit country). In other words, this chapter identifies the factors that can mute the consequences of swings in bargaining power over the pipeline. It also indicates the relevance of the concept of mutual dependency to oil and gas pipeline transit agreements during the operating lives of the pipelines.

6.2 Transit country behaviour – a review of the problems

As discussed in Chapter 2, a number of influences could determine the behaviour of the transit country in terms of capturing rent from the pipeline. As stated in Chapter 1, the literature on this issue is not extensive; however, the works of Stevens (1996, 2000a, 2000b, 2009) comprehensively address these factors, as well as their potential drawbacks. According to Stevens, who analyses the characteristics of transit countries that encourage or discourage further rent seeking during the operation of the pipeline, a number of factors play a significant role in determining whether the transit country will be 'good' or 'bad':

- single versus alternative pipelines;
- political, military, or economic dependence;
- locking the transit country into an off-take agreement;
- the importance of the transit fee to the transit country;
- competition for markets; and
- the importance of foreign investment.

Most of these factors have already been applied to the case studies in Chapter 4. In the case of *alternative pipelines* versus the exporter's dependence on a single line, the potential for the transit country to squeeze is very high if there is no alternative. What an alternative pipeline (or even the possibility of an alternative) does is shift power in favour of the producer country (the exporter), as long as there is no collusion between the alternatives. As was shown in Chapter 4, there were few alternatives to the four case study pipelines. (For the sake of argument, it was suggested that the NREP and WREP constituted significant competition to the BTC pipeline, but the BTC pipeline project was pursued precisely because the NREP and WREP did not have enough capacity, as well as for other reasons, such as environmental concerns.) In the categorisation of 'good' versus 'bad' transit countries, more aggressive rent-capturing behaviour is, therefore, expected from single pipeline cases (e.g. Georgia, Turkey, Benin, Togo, and Cameroon) because there are no feasible and economically viable (and in some cases politically acceptable) alternatives for the exporter, which only serves to increase the bargaining power of the transit countries once the pipeline is in operation.

In terms of *political, military, or economic dependence* of the transit country on the exporter or the consuming market (the importer), or both, the simple argument is that transit countries in such positions tend to be less aggressive. In the WAGP case, for example, the Nigerian and Ghanaian economies are clearly the largest in the region. Peace-keeping efforts in the region (e.g. in Liberia and Sierra Leone) have been largely bankrolled and manned by Nigeria. The threat of military action, as mentioned in Chapter 3, could always serve to check the aggressiveness of the transit country (however unpleasant the prospect would appear). In addition, Benin and Togo are heavily economically dependent on Nigeria. With the BTC, SCP, and Chad–Cameroon projects, economic dependence plays a more significant role (as shown in Chapter 4).

Locking the transit country into dependence on the off-take of oil or gas is another factor that might ensure 'good' behaviour on the part of the transit country, but it could aggravate rather than ease the problem. A strict definition of the transit fee as the off-take volume could ease the problem, but in the case of gas, for example, the transit country could continue to siphon from the pipeline despite its inability to pay, depending on the importance of the pipeline to the exporter (i.e. the exporter's fear of losing the importing market).

The same logic applies in the case of *the importance of the transit fee to the transit country*. Given the bygones rule and the obsolescing bargain, the transit country could continuously seek to renegotiate, because the project will continue to run so long as it meets its operating costs. This is even more the case where the pipeline is operated by the exporter. In this case, the transit country can even capture rent directly from sales.

In terms of *competition for markets*, aggressive or 'bad' behaviour on the part of the transit country could occur if the transit country is a competing source of oil or gas for the destination market (the importer). Once the pipeline has been constructed and is operational, disrupting the flow of oil or gas from the exporter works to the transit country's advantage, either in terms of its own oil or gas being available to the importer using the existing infrastructure or because such disruption may lead to higher prices for its own exports. In the case of the Chad–Cameroon pipeline, for example, the proximity of the pipeline to Cameroon's Logoni Basin poses an opportunity for such behaviour by Cameroon.

If securing *foreign investment* is of great value to the transit country, then it is likely that the transit country will be less aggressive in renegotiations, so as not to scare off potential foreign investors. If, however, the transit country is more selective about the source of such investments or is incapable of attracting foreign investments, it will be easier to squeeze the pipeline for more rent, as the consequence (i.e. the cost of losing current and potential investments) is of little significance to the country.

These factors define the circumstances in which the transit country would act positively or negatively with regard to the transit fee (as far as the exporter is concerned, or in a security of supply context). However, there is no guarantee that these circumstances would elicit specific behaviour from all transit countries. Locking the transit country into off-take volumes, for example, does not prevent it siphoning from the pipeline without payment; nor does the value of the project to the transit country guarantee 'good' behaviour, since the value to the transit country could even be in the additional rent it intends to squeeze from the project. So long as the transit country can optimise rent, and the pipeline continues to meet its operating costs, the threat exists.

The case studies do, however, reveal ways in which the likelihood of 'bad' behaviour by the transit country can be reduced, or even removed; ways in which arbitrary use of the transit country's bargaining power is checked. What follows in the next sections is an attempt to define and classify the ways in which bargaining power shifts can be muted (through what will be referred to as factors of comprehensive mutual dependency in transit oil and gas pipelines).

6.3 Mutual dependency – an attempt at a definition

In the field of sociology and social psychology, Michaels and Wiggins (1976) define mutual dependency as a mutually profitable exchange, such that two parties are partially but unequally dependent on themselves for outcomes or rewards. In conceptualising dependency and power relations in terms of relating one party's rewards to another party's behaviour, Emerson (1969) argues that the more a party's rewards are dependent on the behaviour of another, the less that party's relative power in the exchange relation. Therefore, the dominant power is held by the party that does not depend on the relationship for rewards or outcomes. Emerson notes that a dependency balance (or equal dependency) would discourage the use of power by either party, but further states that unbalanced relations tend to move towards balance over time. In terms of an empirical definition, mutual dependency can be defined as the 'ratio of the sum of the two persons' dependent payoffs to the sum of their dependent and independent payoffs' (Michaels and Wiggins, 1976).

The notion of mutual dependency in the social sciences stems from social exchange theory, which suggests that people weigh the rewards (such as social status and social support) and costs (such as conflict and inequities) of particular behaviours (Janz, 2000). There are several variants of social exchange theory, but the most often cited is the work of Thibaut and Kelley (1959), who state that people also make comparisons with two other standards (the *comparison level* and *the comparison level for alternatives*) when determining the benefits of an exchange situation (in addition to considering the outcomes or rewards). The literature on social exchange theory is quite vast (although quite dated)[1] compared with what is available specifically on the concept of mutual dependency. However, in terms of a general definition of the concept, the gist of mutual dependency is reciprocal cooperation between two or more parties to reap the benefits of this cooperation. The concept also encompasses the minimisation or removal of conflicts that would ordinarily ensue in such relations.

On this basis, a simple definition in the context of the research objective of this book (i.e. analysing the behaviour of transit countries with regard to exercising their bargaining power in transit fee negotiations or renegotiations) would define mutual dependency as a relationship between the principal parties to a pipeline agreement (in this case, the exporter, transit country, and importer) that checks arbitrary use of the bargaining power of the transit country, such that there is less likelihood of a dispute resulting from the transit country trying to extract more rent from the pipeline. The relationship would be influenced by factors in the agreement among the parties, as well as (and more importantly) by factors exogenous to the project, both directly and indirectly related to the pipeline, which influence the decisions of the transit country. The relevance of this concept to this book, as discussed in the introduction to this chapter, stems from the potential problems that could still arise from the transit country's bargaining power in respect of the pipeline project if the solution to the transit fee problem is based on pure bargaining alone.

As the case studies in Chapter 4 and the following sections of this chapter show, the concept of mutual dependency in transit pipeline agreements can be further classified into two:

- endogenous mutual dependency; and
- exogenous mutual dependency.

Endogenous mutual dependency can be defined as mutual dependency among the parties to the pipeline agreement influenced by factors directly related to the agreement itself. The pure bargaining principles employed in the analysis of the case studies fit into this definition. Thus, for example, contributions to meeting project costs, a common interest in the construction of the pipeline, and the relevance of the pipeline to the transit country all count as mutual dependency ingredients endogenous to the pipeline agreement.

Exogenous mutual dependency is an extension of the mutual dependency definition. It expands the scope of mutual dependency to take into account factors outside the pipeline agreement that can also have an impact on the bargaining positions of the parties to the agreement. In discussing the motive behind the four pipeline case studies, Chapter 4 identified economic and political influences (from the US in the case of the BTC and SCP projects; from Nigeria in the case of the WAGP; and the role of the World Bank in the case of the Chad–Cameroon pipeline) which were not directly related to the pipeline agreement, but which could play a key role in redistributing bargaining power away from the transit country during the operation of the pipeline. The following section sheds more light on exogenous mutual dependency; it identifies mutual dependency ingredients which are essentially exogenous to the pipeline project but have a significant bearing on the transit country's decision to rent-squeeze or not.

6.4 The ingredients of mutual dependency

In light of the factors that determine 'good' versus 'bad' transit countries, it is clearly a struggle to counter-balance the bargaining power of the transit country. However, the history and development of the case study pipelines reveal certain key aspects that could augment the factors mentioned in section 6.2, such that the potential for disruptions due to renegotiation of transit terms is reduced or removed. This section identifies four key factors, which are discussed in order:

1. the role of strategic investments in the transit country;
2. the role of complementary projects;
3. the role of international relations; and
4. the role economic and political alliances (strategic alliances).

6.4.1 The role of strategic investments in the transit country

In their analysis of the shift in the balance of power along the vertical supply chain for (Eurasian) natural gas, Hubert and Ikonnikova (2003) argue that the power shift is a function of the architecture of the transmission network and that the balance of power can, therefore, be altered through strategic investment in pipeline capacity. They conclude that the power of a player should increase as the player becomes more 'important' to the other players, depending on the former's control over the gas fields and transport routes. This suggests strategic investment in pipeline capacity to increase the importance of both the exporter and the destination country to the transit country. The importance of these players can be measured by looking at the contributions each player can make to the various possible coalitions.[2]

Example

> Assume three participating countries in a transit pipeline agreement: country **A** is the exporter (or producer country), country **B** is the transit country, and country **C** is the importer (or destination country). On the basis of the findings of Hubert and Ikonnikova (2003), and the discussions in Chapters 3 and 4, country **B** tends to possess more bargaining power once the pipeline has been built and put into operation, and can as a result seek to renegotiate transit terms during the operation of the pipeline. What the work of Hubert and Ikonnikova advocates is strategic investment by **A** or **C** in the capacities of other pipeline routes (say through transit countries **D** and **E**) – assuming there is an existing pipeline transmission network – such that there are alternatives to the A–B–C route; this

enhances their value to country B in the pipeline arrangement, and checks the bargaining power of country B.

Hubert and Ikonnikova's analysis of the Eurasian supply chain for natural gas takes Russia's options for exporting gas to Western European markets as an example. It identifies a potential threat in such transit countries as Ukraine, and makes reference to Russia's attempts at diversification (its agreements with Belarus and Poland – and, more recently, with China) as an attempt to check this threat. This has been argued to encourage Russia to develop other fields (Gazprom), as well as to invest in increasing the capacity of other pipelines, to neutralise the bargaining power of Ukraine. An example given is the plan for a twin pipeline with a capacity of 60 billion cubic metres per annum running north–south through Belorussia, Poland, and Slovakia (known as the Yamal 2 pipeline).

The problem with the above analysis is that some of its assumptions make it difficult to apply to the pipeline cases studied here, which leads to question marks concerning its overall applicability to pipelines. This is essentially a problem of scope. First, the analysis assumes that the parties are risk neutral, and, therefore, equally interested in expected payoffs. It is reasonable to argue that this is not the case with the pure transit country, as it can afford to be less risk averse given the fact that it will not be worse off without the pipeline. Second, the analysis assumes that coalitions could be formed by the exporter and importer, and indeed the transit country (or countries); this carries an implicit assumption of alternative pipeline routes (if an attempt is to be made to apply this analysis to oil pipelines) or a functioning gas pipeline network that would encourage such coalitions. This is not the case with the WAGP, Chad–Cameroon, BTC, and SCP projects. Strategic investment in the capacities of other pipelines to strengthen the bargaining power of the exporter and the destination market obviously cannot work without such pipelines.

The idea of strategic investment as a tool to neutralise the bargaining power of the transit country is not, however, totally useless to this book. If the scope is broadened to include investments other than pipeline investments that have direct or indirect significance to the transit country, the concept becomes more suited to forging mutual dependency among the parties. The Russian case used above is an example of strategic investments yielding positive effects for the exporter (or the importer), but it cannot be used as an example for a more broadened scope, which would include investments outside the pipeline agreement.

The original features of the proposed Iran–Pakistan–India (IPI) gas pipeline shed more light on the nature of strategic investments that could serve to enhance the bargaining positions of the exporter and the importer, as well as ensure that the transit country does not arbitrarily seek renegotiation of transit terms.[3]

Under the proposal, India was expected to draw on two-thirds of the gas, while Pakistan would get the remaining one-third. Of the three options for transporting gas from Iran said to have been considered by India (LNG, deep-sea pipeline, and on-land pipeline from Iran's South Pars field), the IPI pipeline was claimed to work out four times cheaper than the other alternatives, even after factoring in a transit fee for Pakistan.

The possible scenarios for an agreement to be reached were to be determined by a number of key factors, and are listed here because of their relevance to this discussion:

- 2.5 million tons of gas per annum at *half* the international price, payable *after* delivery, which would make the project more attractive to India;
- Iran's suggestion that an international consortium of bankers and oil companies would own and operate the pipeline so that India and Pakistan did not need to deal directly with each other;
- locating the spigots (pipeline taps) only in Iran and India to prevent Pakistan turning off supply without actually blowing up sections of the pipeline, thereby hurting its own supplies; and
- a proposed electricity–gas exchange in which India would enter into power supply contracts with Pakistan (i.e. the gas would be piped to India to fuel power plants that would sell power back to Pakistan); this would have ensured Pakistan's dependence on India, as well as counter-balancing the risks of Pakistan cutting off gas supply.

If any or all of the above terms were eventually to apply to the IPI pipeline (if, indeed, this pipeline materialises at all), then it is argued that the bargaining positions of Iran and India would be significantly enhanced compared with Pakistan's position. There are, however, risks of disruption of the pipeline by terrorist groups. For example, two gas pipelines in Pakistan were blown up to send a message of disapproval of the proposed IPI pipeline in 2005.

In the examples above, the factors that could serve to check the bargaining power of the transit country are endogenous to the pipeline.

There is also an exogenous side to the strategic investment argument, as demonstrated in the following sub-sections.

The BTC and SCP projects: the role of FDI in Georgia and Turkey

As discussed in Chapter 4, Western interests in the BTC and SCP projects are not only obvious, but also huge from Georgia's perspective. Investments in Georgia (however apparently unrelated to these two pipelines running through its territory) play a key role in the country's bargaining position. Research into the IFC's investments in Georgia between 1998 and 2005 reveals significant investments in sectors of key relevance to the country.

The primary objective of the IFC's assistance to Georgia is to provide support for the development of the housing finance market and to reach small and medium-sized enterprises. The IFC has provided loans to Georgia in the power sector, the manufacturing sector, and the oil and gas sectors.

The International Development Association and the IFC prepared a country partnership strategy for Georgia (IDA/IFC, 2005) in which a detailed development assistance programme is documented. In this document, significant focus is put on the following areas (in no specific order of importance):

- macro-stability, budget assistance, and debt amelioration;
- agriculture/rural development;
- oil and gas;
- transport/telecommunications;
- power;
- education;
- health;
- the environment;
- water supply and municipal development;
- public sector reform;
- community empowerment; and
- fiscal/budgetary management.

In an attempt to liberalise the country's economy, the Turkish government introduced a flexible foreign investment policy, backed by legislation, to provide a secure environment for foreign capital. The main features of this legislation include a national treatment principle, minimum capital requirements, no limits on participation ratios, and investment incentives such as exemptions from taxes, duties, and fees.

IFC investments in Turkey cover a range of sectors, including:

- health (hospital expansion projects);
- the BTC pipeline;
- manufacturing (steel and pipe production company restructuring);
- commerce and downstream business (Kusadasi Port expansion, retail distribution of petroleum products); and
- the financial sector (asset securitisation, redevelopment, lending to small and medium-sized entities, etc.).

In October 2003, the World Bank and the IFC produced a country assistance strategy for Turkey in which strategic investment was geared to fit into Turkey's vision and economic priorities. The strategy is structured around four development themes:

- sound macroeconomics and governance;
- human and social development;
- an attractive business climate and business knowledge; and
- the environment.

In Georgia, the sectors identified above have received financial assistance from donors and partners ranging from the World Bank to the UN and UK agencies. The aim behind listing all the sectors on which there is significant focus in the country partnership strategy is to depict the vastness of the Western community's dedication to the developmental needs of Georgia. Virtually every key sector in the Georgian economic framework benefits from one form of financial assistance or another. This has obvious implications of obligation on the part of Georgia. Moreover, the timing of these investments can be argued to be no coincidence. On the basis of the data available from the IFC on sponsored projects in Georgia (there was significant increase in financing from 1998, which follows a similar timeline to the BTC and SCP projects) and the release date of the country partnership strategy (October 2005), it is easy to assume that these financial packages are strategic investments in Georgia, such that Georgia will seriously consider returning favours to its donors, one of these being 'good' behaviour as far as the BTC and SCP projects are concerned. The argument is similar for Turkey, where the interest in strategic investment is far larger than in Georgia: these investments can be argued to create an implied obligation on the part of Turkey to its donors, especially with respect to huge projects such as the BTC.

The WAGP project: the role of FDI in Benin and Togo

In the case of the WAGP, there have been significant investments in Benin and Togo to make similar assertions about what is expected of them as in the case of Georgia and Turkey. The World Bank and the IFC are also involved in key projects in Benin and Togo, playing investment roles in such sectors as finance and insurance, industrial and consumer products, health, information, manufacturing, and power. The IFC's first microfinance investment project in Benin commenced in 2002, with the broad objective of generating new income and improving living standards. Specifically, the microfinance investment project is aimed at helping low-income businesses that lack access to credit to gain access to finance for start-up or expansion.

The low levels of economic performance in these two countries were discussed in Chapter 5, and it is obvious that their economic development strategies will require significant financial aid for their major projects. Their dependence on donors can, therefore, be argued to ensure that they meet obligations with respect to other agreements reached with these donors.

It is important, however, to point out that Benin and Togo have not waived their rights to charge a fee in addition to the gas off-take.[4]

The Chad–Cameroon oil pipeline project: the role of strategic investments in Cameroon

The IFC has been committed to financing projects in Cameroon since 1956. The strategic priorities for which the IFC has provided assistance in the Cameroonian case are as follows:

- an improved business climate and dialogue between the public and private sectors;
- technical, managerial, and financial support to micro, small, and medium-sized enterprises;
- private sector participation and the financing of infrastructure projects; and
- financial and technical assistance to microfinance activities and small business subcontractors of the Chad–Cameroon pipeline project.

It should be noted at this point that there is no direct link between the dependencies of these countries outside the pipeline projects and their potential behaviour regarding the secure transit of oil or gas via their territories. It should also be observed that with the exception of Turkey, all the transit countries have two common characteristics – a poor economy and weak international presence. These characteristics have implications for the developments discussed in this chapter, as will be shown in the following sections (especially section 6.4.3).

6.4.2 The role of complementary projects

In the four pipeline case studies, Georgia and Turkey are the only transit countries with separate oil and gas pipelines passing through their territories. The argument here is that complementary pipelines could positively affect the transit country's behaviour (i.e. reduce the bargaining power of the transit country), on two conditions:

- that there are close ties between the two pipelines in terms of motive, ownership, operation, and financing; and
- that at least one pipeline has value to the transit country.

In the case of the BTC and SCP projects, on the basis of the available information surveyed in Chapter 4, the sources of participation and sponsorship are essentially the same. In terms of equity participation, BP has the largest share (30% in the BTC pipeline, 25% in the SCP). In terms of country participation, Western European countries dominate both projects, and, most importantly, the sources of finance for the projects are similar (with BP and the EBRD as the leading contributors). In terms of motive, the argument for a diversified source of Caspian energy (i.e. other than via Russian routes in particular) is a shared objective behind the construction of the pipelines (as detailed in Chapter 4). The large power generation and industrial gas consumer segment in Georgia is of strategic value to the Georgian economy. Georgia's dependence on Shah Deniz gas was discussed in detail in Chapter 4, where the following key benefits to its domestic gas market were pointed out:

- the potential to diversify its sources of gas;
- the potential to ease its gas feedstock problem, which could enable it to meet up to 20% of its winter peak needs (or 80% of Tbilisi's peak needs); which, therefore, implies
- a competitive gas market, resulting in reduced prices in the long term.

Georgia is clearly dependent on SCP gas. Therefore, it can be suggested that clauses could (perhaps *easily*) be incorporated into the pipeline agreements such that interruption of one pipeline for transit fee renegotiations by the transit country could affect its earnings from the other. This could, however, be affected by the amount of rent the transit country receives from the other pipeline: it is possible that this rent could be sacrificed in pursuit of more rent from the other pipeline with the larger 'prize'. If a clear and significant dependence on at least one of the pipelines can be determined (as in the Georgian case for gas), it becomes difficult for the transit country to disrupt the complementary pipeline over renegotiation of transit terms if a link between the lines (motive, finance, participation) exists.

The WAGP and Chad–Cameroon projects do not have complementary pipelines, and thus have not been included in this analysis. However, given the same conditions of motive, ownership, and financing, and the value of at least one pipeline to the transit countries, one could argue the same effects as in the BTC and SCP cases.

6.4.3 The role of international relations

The argument here is based simply on the transit country's aim of international presence or association. So long as the transit country attaches importance to its membership of an international institution, it will, *ceteris paribus*, avoid any threat to that membership. Turkey's position with respect to the BTC pipeline is a useful example, in this case, of how international relations could influence the behaviour of the transit country in either direction. In Chapter 4, in the discussion of whether Turkey would exercise its bargaining power in relation to the BTC pipeline, it was stated that three key factors would influence its decision:

- strong US interest in the pipeline;
- Turkey's vital role in the East–West Energy Corridor; and most importantly (for the purpose of this discussion)
- developments affecting its ambition (or otherwise) to accede to the EU.

Georgia bases its objective of gaining the international limelight on its involvement in the BTC and SCP projects. In his speech at the Washington Business Forum in 2003, the president of the Georgian International Oil Corporation, Giorgi Chanturia, said that the most important milestone in the history of the statehood of Azerbaijan,

Turkey, and Georgia is the realisation of the pipeline projects. Georgia is certain that the actualisation of these two projects will bring countries in the region into the public eye of the West's geo-political and geo-strategic interests. In terms of its relationship with the US, Georgia felt indebted to the US administration and the American nation for what Chanturia (2003) termed 'unwavering support'. It can, therefore, be argued that Georgian commitment to the two pipeline projects (in specific terms) and its commitment to the East–West Energy Corridor (in broader terms) go a long way in demonstrating a mutual dependency between Georgia and the principal parties to the pipeline projects.

In the case of the WAGP, there is huge political and socio-economic dependence of Benin and Togo on ECOWAS. ECOWAS serves as the socio-economic 'regulator' in the sub-region, and the pipeline is key to the promotion of economic integration in such fields of endeavour as natural resources, energy, industry, transport, telecommunications, commerce, and monetary and financial matters. It would, therefore, also not be in the interests of these transit countries to disrupt the pipeline by renegotiating the transit agreements, as this could seriously affect their relationship with the regional institution.

The Benin Republic attaches great importance to its international visibility, especially through its membership of 48 international institutions, notable among which are the African Union, ECOWAS, the UN, the International Atomic Energy Agency, the International Bank for Reconstruction and Development, the IMF, the G-77, the World Health Organization, and the United Nations Educational, Scientific, and Cultural Organization. Togo also places value on such international presence and currently enjoys membership of 46 international institutions, among which the World Bank and the IMF have been its economic supporters. In terms of international association within the continent, Nigeria plays a key role in terms of economic and political influence (especially through the African Union and ECOWAS), and is thus a vital ally for Benin and Togo. Benin and Togo currently enjoy the following benefits from membership of ECOWAS and the African Union:

- growing volumes of intra-trade (or inter-state trade);
- the encouragement of small and medium-sized entities; and
- free movement of goods and people within the ECOWAS region (this is, however, currently relatively restricted compared with other ECOWAS members).

Another factor that can be said to play a key role in the dependency relationship between Benin, Togo, and the two key African institutions to which they are affiliated is that the WAGP project is itself wholly supported by ECOWAS. This implies that disruption of the pipeline for any reason (renegotiation of transit terms in this case) would be viewed as opposing the principles and objectives of ECOWAS. In addition, Benin and Togo are yet to be included on the list of countries that print and operate the ECOWAS travel certificate. Again, such dependency on the part of Benin and Togo as transit countries can positively affect the bargaining positions of the principal parties to the WAGP.

6.4.4 The role of economic and political alliances (strategic alliances)

It has already been stated that even healthy economic and political relationships between the party states to a pipeline agreement do not necessarily check the transit country's bargaining power, simply because of the obsolescing bargain. When a transit country has such potential bargaining chips as rent calculations being based on a fixed price for oil or the right to charge a fee in addition to off-take volumes of gas, there remains a temptation to squeeze for as long as there is rent from which to squeeze.

Broadening the scope of economic and political alliances beyond the party states to the project to include relationships of economic and political dependence of the transit country on the larger international community slightly alters the playing field. In the case of the BTC and SCP projects, for example, there is evidence from the case study chapter to suggest a strong alliance between the US and Georgia. In terms of security of supply, the Caspian region offers an alternative source of petroleum for Western markets. Section 4.3.1 mentioned the argument that an oil pipeline from Azerbaijan to Turkey would be of potentially great benefit to many US and regional policy goals. It also discussed that the general argument for the role of the East–West Energy Corridor concerns the expectation that the corridor will play a significant role in achieving the US geo-political and geo-strategic objective of unlocking Caspian energy reserves and securing the supply of this energy to Western markets. This objective has been further reinforced by suggestions that there would be a significant increase in world dependency on the Middle East. Georgia views the East–West Energy Corridor as having immense potential for development. Tbilisi expects that the corridor will ensure peaceful and secure development in southwest Europe.

In the WAGP and Chad–Cameroon projects, there is also evidence to suggest a dependency relationship between the transit countries and the larger community. A cursory look at the politics and economics of Benin, Togo, and Cameroon reveals the following;

- Togo has historically enjoyed positive relations within the African continent, predominantly with former Zairian leader Mobutu Sese Seko.
- Previously sour relations between Togo and Ghana (which was repeatedly accused of sponsoring coups in Togo) improved following the election of President Kufuor of Ghana in 2000.
- Togo has facilitated bilateral trade with Ghana, and inter-regional trade with the rest of the ECOWAS region since 2002. (Togo's imports from Ghana account for 26% of total imports; and the country exports to Nigeria, Benin, and Ghana).
- Benin's foreign policy is limited, in terms of priority, to the West African sub-region, primarily Nigeria, Niger, Togo, and Burkina Faso.
- Benin is heavily economically dependent on Nigeria, so it tends to avoid disagreement with its relatively more powerful ally.
- Benin's influence within the West African sub-region, and the wider African region, is limited.
- Cameroon's foreign policy can be described as conservative, non-confrontational, moderate, and pro-West. This arguably has positive impacts on the country's trade relations with Chad, the other members of the Central African Economic and Monetary Union (Equatorial Guinea, Gabon, Congo, and the Central African Republic), the EU, the US, and China.

It is, therefore, argued that so long as the transit country values its economic and political dependence on other countries that are also linked to the pipeline (as well as other international alliances that affect its economic and political standing), by way of finance, ownership, or as a market for the hydrocarbon transported via the pipeline, its bargaining power can be checked. So long as the economic and political costs of jeopardising such alliances outweigh the benefits from arbitrary renegotiation of transit terms, the transit country will be reluctant, *ceteris paribus*, to disrupt the pipeline for more rent.

6.5 Conclusion

What this chapter has shown, on the basis of the case studies analysed in Chapter 4, is that in the context of this book, comprehensive mutual dependency (endogenous and exogenous) is an economic and political relationship between the exporter, transit country, and importer that reduces or limits the transit country's power to arbitrarily renegotiate transit terms. The four main case studies used in this research have shown that strategic investment in the transit country can work to commit the transit country to other obligations. The original features of the proposed IPI pipeline provide a remarkable example of the direction which future transit oil and gas pipeline agreements could take (via the *endogenous mutual dependency* route). Such agreements would reduce the need for huge investments in other sectors of the transit country's economy. Complementary pipelines will have a positive effect only if there are close links between the pipelines and if the value of at least one of the pipelines is significant to the transit country. Finally, international institutions, depending on how much influence they have over the transit country, can also have a positive effect, as can positive economic and political alliances between the transit country and the principal parties to the pipeline.

Although the ingredients of *exogenous mutual dependency* discussed in this chapter have been shown apparently to limit the bargaining power of the transit country when the pipeline is operational, it is important to note that these factors are heavily dependent on the value attached by the transit country to each of them. A transit country not keen on foreign investment might not be concerned about the consequences of squeezing the pipeline for more rent, but one that attaches value to its international political and economic stance is likely not to disrupt the pipeline for a rent squeeze.

In the context of the analysis so far, and for the sake of simplicity, there are two possible bargaining power outcomes during the operation of the pipeline: the transit country might exercise its bargaining power and squeeze more rent from the pipeline or it might not exercise such power (either because it is not in a bargaining position to do so or because it chooses not to do so). The objective throughout this book, however, has been to identify the factors that could yield a situation in which transit countries are not able to arbitrarily rent-squeeze. What the analysis and application of the principles of bargaining in Chapters

3 and 4 has shown is that bargaining alone could increase the likelihood of a balanced bargaining power scenario for all the principal parties to the pipeline. However, the bargaining conditions (i.e. common trade interests, off-taking as compensation to the transit country, the economic and geo-political value of the pipeline to the transit country, competition for markets, and cost-sharing) can be argued to be double-edged swords, in that they could (individually or as a whole) also serve as reasons for the transit country to seek more rent (or off-take) from the pipeline. With the exception of the dependency example from the IPI pipeline, this outcome can also be termed the *endogenous mutual dependency outcome*.

The ingredients of mutual dependency – as demonstrated in this chapter, on the basis of the four case studies and the IPI pipeline – tend to have a more stable effect in terms of confining the transit country's bargaining power. One or any combination of these factors can prevent out-of-agreement renegotiation of transit terms by the transit country, thus reducing the likelihood of disruption to the operation of the pipeline. This is represented in Figure 6.1 by the grey boxes. This, in turn, can be termed the *exogenous mutual dependency outcome*.

An arguably more restrictive (perhaps desirable) solution to the threat of transit fee renegotiation could be 'the bargaining outcome' (as depicted in Figure 6.1) resulting from the fusion of the two sets of factors that yield the endogenous bargaining outcome and the exogenous bargaining situation outcome (i.e. endogenous *plus* exogenous mutual dependency). This, I argue, is the point of comprehensive mutual dependency between the exporter, the transit country, and the importer. More importantly, the cases used in this book have shown that such confinement of transit country bargaining power is practicable.

It should be noted at this point that this chapter has not argued whether the significant bargaining power of the transit country (and thus the task of seeking means of reducing this power) is a good thing or otherwise. There can be genuine arguments from each of the principal parties to the agreement as to why one should possess more power than the others. However, what this chapter (and indeed the entire book) has done is take a perspective based on the negative impacts of interruptions on the costs of the pipeline (to producer/exporter, transit country, and importer alike) and thus the objective of seeking a solution to reduce such costs.

Notes

1. E.g. the seminal works of Blau (1964), Homans (1958), Thibaut and Kelley (1959), and Heath (1976). A detailed discussion and review of these works can be found in Emerson (1976). There are trickles of recent work using the theory, e.g. Lawler (2001).
2. The method of measurement uses a Shapley value function that derives the payoffs of the players from the fundamentals of the problem. How this function is derived is complex, and not of importance to the discussion here, and is, therefore, not pursued further; it is explained in the Hubert and Ikonnikova article. Ikonnikova (2006) suggests a partition function form solution instead of the Shapley function, but this approach is flawed in terms of its applicability given the possibility of coalitions being formed by transit countries. See also Kargin (2003).
3. This pipeline project has not been used as one of the main case studies in this book owing especially to the unreliability of available data sources.
4. An informal discussion with an attorney at the Nigerian National Petroleum Corporation revealed concern over the potential problems this could cause in the future.

7
Concluding Remarks

Abstract: *The findings from this book suggest that while, at the time of signing the initial agreement, the parties must have found its terms to possess the characteristics of reasonableness, objectiveness, transparency, and non-discrimination, their post-construction behaviour shows that such characteristics are not enduring. Evidence from the case studies shows endogenous mutual dependency conditions to be prevalent in the pipeline projects. It becomes necessary to widen the scope of the definition of mutual dependency, as this enables the identification of exogenous conditions, which will inevitably encompass issues pertaining to politics and international relations. These, in combination with the endogenous conditions, provide a comprehensive mutual dependency bargaining framework that will serve as a deterrent to arbitrary rent-seeking by the transit country.*

Omonbude, Ekpen James. *Cross-border Oil and Gas Pipelines and the Role of the Transit Country: Economics, Challenges, and Solutions.* Basingstoke: Palgrave Macmillan, 2013. DOI: 10.1057/9781137274526.

7.1 The research questions

This book set out to answer two questions of importance to the security of supply of oil and gas transported via transit pipelines. The first question centred on the scope of definition for pipeline agreements involving transit. Is there any such animal as a transit pipeline agreement that – as far as all the principal parties are concerned – is reasonable, is objective, is transparent, and is non-discriminatory? Is there any such agreement that has in-built mechanisms to prevent future disputes, especially the kind that could arise from the transit country's dissatisfaction with the transit terms during the operation of the pipeline?

The findings from this book suggest that while, at the time of signing the initial agreement, the parties must have found its terms to possess the above characteristics, their post-construction behaviour shows that such characteristics are not enduring. The economics of rent and the economics of bargaining make this clear (see Chapters 2 and 3). So long as the pipeline project earns supernormal profit, so long as bargaining power shifts to the transit country once the pipeline has been built and is operational, and so long as the transit country is aware of this shift and has factored in the various constraints to optimising its rent, the transit country *can* disrupt the pipeline in pursuit of a larger share of that rent. This is both a theoretical and a practical possibility. It is practically possible simply because even when dispute settlement mechanisms and a means of international legal recourse are available to aggrieved parties, the potential of disruption by the transit country is not completely removed. If the potential value to the transit country of reneging on the pipeline agreement is higher than the punitive consequences of breaching any such agreement, the transit country can disrupt the pipeline. The problems with this situation are the huge costs of disruption to the exporter and the security of supply implications for the importer.

The second (and main) question centred on how the consequences of shifts in bargaining power among parties to a cross-border oil and gas pipeline can be muted (if they can be at muted at all). The book attempted to develop a workable framework that could help check the abuse of bargaining power by the parties to the pipeline. The case studies selected for this book revealed a number of important political and diplomatic factors endogenous and exogenous to the pipeline projects that demonstrated two things: one, that it is indeed possible to check the swing in bargaining power to the transit country to prevent arbitrary rent-seeking during the operation of the pipeline, and, two, how such swings can be checked. A framework of comprehensive mutual dependency was thus developed as *a solution* to the transit fee dispute puzzle. The argument is that the potential for pipeline disruptions by the transit country during the operation of the project is a bargaining problem, and therefore the solution must be derived from the basic principles of bargaining.

7.2 A comprehensive mutual dependency solution

Evidence from the case studies at first shows endogenous mutual dependency conditions to be prevalent in the pipeline projects. This conclusion is based on the bargaining principles outlined in Chapter 3 and the useful elements of each project's history and development analysed in Chapter 4. The motive behind the project has been identified as a significant part of the mutual dependency framework. Understanding what each party to the agreement intends to gain from the pipeline is useful to the mutual dependency framework, as this influences its behaviour during the operation of the pipeline, especially when bargaining power shifts to the transit country.

Other factors, such as participation in cost-sharing and the availability of alternative pipeline routes, also play key roles in regulating transit country behaviour during the operation of the pipeline. Participation in cost-sharing, for example, could check aggressive behaviour by the transit country simply because it is an active participant or stakeholder in the pipeline. In the same vein, alternative pipeline routes also provide a useful solution to the bargaining problem, so long as the alternative pipeline is viable.

However, it is reasonable to argue that the endogenous conditions are insufficient deterrents to arbitrary rent-seeking behaviour by the transit country. The case studies, for example, have shown that actual cost participation by the transit country can be minimal, and, therefore, arguably not enough to restrict aggressive rent-seeking behaviour. In some fiscal arrangements (e.g. production-sharing contracts), the government concern is carried through the project, despite some stake ownership or 'cost participation'. This will not deter the transit country from pursuing a higher transit fee if it feels there is more to gain. However, if the arrangement contains elements of progressivity, such that transit fee increases or decreases are linked to upward or downward movements in the value of the project, then there is a likelihood of reduced temptation to renegotiate fiscal and commercial terms. Alternative pipeline routes will reduce the transit country's bargaining power. However, this has been shown to be uneconomical.

Therefore, it becomes necessary to widen the scope of the definition of mutual dependency, as this enables the identification of exogenous conditions, which will inevitably encompass issues pertaining to politics and international relations. These, in combination with the endogenous conditions, provide a comprehensive mutual dependency bargaining framework that will serve as a deterrent to arbitrary rent-seeking by the transit country.

7.3   The role of the Energy Charter

The Energy Charter Secretariat has not acknowledged the consequences of shifts in bargaining power between the exporter and the transit country, as is evident in the provisions of both the ECT and the proposed ECTP. Admittedly it will be difficult to set general boundaries for what it is reasonable for the transit country to collect as a transit fee in a pipeline agreement. This may still have to be done on a case-by-case basis, given the peculiarities of each project. However, the potential for arbitrary pipeline disruption due to a rent squeeze is not a case-by-case problem, although this has not been acknowledged in the ECT, nor has it been acknowledged in the literature. The question of why it has not been acknowledged in the ECT or the draft Protocol thus arises. There are two main reasons for this:

- the assumption or expectation that transit countries will adhere to the principle of freedom of transit (as stated in GATT Article V); and
- the expectation that transit countries will not seek fees that are not justified by costs or that are discriminatory.

As highlighted in Chapter 5, discussions with a senior member of the Energy Charter Secretariat revealed some recognition of the potentially powerful position of the transit country. The expectation from this would be that provisions would be made to tackle problems emanating from transit countries exploiting this position, but this is not yet the case.

7.4 Limitations of the study

7.4.1 Approach of the book

This book has attempted to build up to the mutual dependency framework as *a solution* to the transit fee dispute puzzle. There is little by way of literature specifically on transit issues in cross-border oil and gas pipeline projects, and this has influenced the theoretical framework for this study. A previous study (the ESMAP paper referred to in Chapters 1, 3, and 5) employed the framework of pipeline economics and hinted that the problem with transit pipelines is essentially a bargaining one. This logically shaped the direction of the theoretical build-up of this book towards the analysis and application of bargaining principles to cross-border oil and gas pipelines involving transit. The endogenous mutual dependency factors identified from the case studies (e.g. cost-sharing, off-taking by the transit country, and economic relevance of the transit fee) were derived from the bargaining principles used to analyse these cases (e.g. inside and outside options, commitment tactics, and risk aversion). Further examination of the pipeline cases in a much wider context of influential factors outside the actual pipeline agreement and fundamental pipeline economics revealed the exogenous mutual dependency extension to the bargaining solution suggested at the end of Chapter 6, which borrows on the concept of mutual dependency from the literature on social exchange theory.

The question of whether another approach could be taken depends essentially on the research questions. If the research question was to determine the basis for apportioning the rent from oil and gas transit pipelines, for example (a problem well highlighted in the course of this research), then this approach would not be suitable perhaps. Focus could then be shifted to optimum rent-sharing mechanisms and possibly an analysis of rate-of-return systems that could be applicable to such cases; or an examination of key fiscal (and perhaps technical) aspects of cross-border oil and gas pipelines involving transit through one or more countries; or a more quantitative, game-theoretic approach.

7.4.2 Morality of the mutual dependency solution

The question of the morality of endogenous and exogenous (especially exogenous) mutual dependency as a solution to the bargaining problem cannot be avoided. It is not unreasonable to argue that the exogenous mutual dependency ingredients identified in this book are a form of blackmail of the transit country. If the findings of the first part of the book suggest pure bargaining as the solution to the rent-sharing problem with respect to transit pipelines, why can this solution not be ideal? (That is, why not leave it to naked bargaining, such that the party with more bargaining power determines its share of the rent?) This line of debate can continue indefinitely, and is not within the scope of this book. The angle taken for this book was informed by the implications of pipeline disruptions (specific to transit terms) for security of supply, and the costs incurred by exporter and importer alike.

7.4.3 Foreign investment by the transit country

In the identification and analysis of FDI as one of the exogenous mutual dependency ingredients, only one angle was considered: foreign investments into the transit country and the effects of such investments on bargaining positions. The reverse situation was not considered, namely investments by the transit country in the exporter country, or the importer country, or other countries with significant influence over the pipeline (e.g. other project sponsors). As discussed in section 6.4.1, this would involve strategic investments in such sectors as finance, commerce, or even energy. It is not unreasonable, for example, to suggest there is a strategic significance to Saudi investments in the US, and that mutual dependency has been enhanced by such relationships. This book did not investigate the case studies for evidence of such investments by the transit countries (strategic or coincidental) that could have an effect on its bargaining position. However, it is not unreasonable to hypothesise a reverse strategic investment (i.e. by the transit country in the exporter country or the importer country) as another exogenous mutual dependency ingredient in the solution to the bargaining problem.

7.5 Further research

As discussed in Chapters 1 and 4, there were limits to the choice and number of pipeline projects selected as case studies for this research. The pipeline cases used in this book are fairly recent; therefore, it is difficult at this stage to predict their long-term success or failure in absolute terms. Assuming transparency of available pipeline data, an ideal course of future analysis would be to conduct similar research on other transit oil and gas pipelines using the mutual dependency framework determined in this book. This book has shown that it is possible to examine the consequences of shifts in bargaining power in the transit pipeline context taking into consideration the elements of the comprehensive mutual dependency framework that has been developed. This could – in addition to testing the general applicability of the findings from this book – identify further ingredients of endogenous and exogenous mutual dependency.

This book did not explore the legal issues surrounding the disputes that could arise from pipeline disruptions as a result of arbitrary renegotiation of transit terms. This was evident in the exploration of the major provisions of the ECT, the draft ECTP, and the dispute settlement mechanisms in GATT and WTO law. This was deliberate. The main focus of this book was the economic and political behaviour of the countries involved in a transit pipeline agreement, hence the narrowed methodology of pipeline economics and bargaining principles (as well as the brief review of mutual dependency as defined in social exchange theory). In light of the findings from this book, another interesting angle for future research would be to assess the implications of implementing a framework of mutual dependency in terms of provisions in international law for energy in transit across international borders, as well as dispute settlement (and even dispute *prevention*) mechanisms within the ECT, the Transit Protocol (if and when it comes into force), and the International Centre for Settlement of Investment Disputes.

Finally, it would be interesting to see how (or whether) the exogenous mutual dependency conditions could be factored into transit pipeline agreements. The four case studies show that although there is clear evidence of external factors that serve to check aggressive rent-seeking behaviour on the part of the transit country, these external factors are not documented in the host government and inter-governmental agreements governing the pipelines. It is not unreasonable to ask whether these factors should be incorporated into the agreements at all, or to suggest that the *status quo* is fine, but the examination of other pipeline cases could reveal the absence of such exogenous conditions. Therefore, it would be interesting to investigate the mutual dependency framework further in the context of whether or not to incorporate the exogenous factors into the pipeline agreement, and also in terms of the implications of such incorporation (economic, legal, and political).

References

Acikalin, N.S. (2003) 'Energy Corridor: Turkey', paper presented at the IEA roundtable on Caspian oil and gas scenarios, Florence, Italy, 14–15 April. http://www.iea.org/work/2003/caspian/acikalin.pdf.

Adelman, M.A. (2002) *The Genie out of the Bottle: World Oil since 1970*. Cambridge, MA: MIT Press.

Agreement on the Exploration, Development and Production Sharing for the Shah Deniz Prospective Area in the Azerbaijan Sector of the Caspian Sea, between The State Oil Company of the Azerbaijan Republic *et al.* (1996).

Alexander's Gas and Oil Connections (2004) 'Saudi Arabia Denies Political Motive in Gas Deals', *Alexander's Gas and Oil Connections* 9(6), March. http://www.gasandoil.com/news/2004/03/cnm41244.

Alexander's Gas and Oil Connections (2001) 'Shah Deniz Gas Pipeline Expected to Cost US$1bn', *News and Trends* 6(14), 31 July.

Avidan, A.A. (1997) 'Lowering the Cost of LNG Delivery: Impact of Technology', paper presented at the 15th World Petroleum Congress, Beijing, October 12–17.

Bamberger, C., J. Linehan, and T. Waelde (2000) 'The Energy Charter Treaty in 2000: In a New Phase', *CEPMLP Online Journal* 7(1).

Baran, Z. (2005) 'The Baku-Tbilisi-Ceyhan Pipeline: Implications for Turkey', working paper, Central Asia-Caucasus Institute, Silk Road Studies Program. http://www.silkroadstudies.org/BTC_6.pdf.

BBC (British Broadcasting Corporation) (n.d.), 'Country Profile: Georgia'. http://news.bbc.co.uk/2/hi/europe/country_profiles/1102477.stm.

Besanko, D.A. and R. R. Braeutigam (2002) *Microeconomics: An Integrated Approach*. New York: Wiley.

Billmeier, A., J. Dunn, and B. Van Selm (2004) 'In the Pipeline: Georgia's Oil and Gas Transit Revenues', IMF Working Paper WP/04/209, Washington DC, November.

Blau, P.M. (1964) *Exchange and Power in Social Power*. New York: Wiley.

BP (British Petroleum) (2002–2011). *BP Statistical Review of World Energy*.

Brennan, T.J. (2002) 'Preventing Monopoly or Discouraging Competition? The Perils of Price-Cost Tests for Market Power in Electricity', Resources for the Future, Washington DC, Discussion Paper 02-50.

Brito, D. and J. Rosellon (2001) 'Timing of Investment in LPG Pipelines in Mexico', research paper, Department of Economics, Rice University, Houston, TX.

Caspian Development and Export (2003) *Economic, Social and Environmental Overview of the Southern Caspian Oil and Gas Projects*, briefing paper, February.

Chanturia, G. (2003), 'The East-West Energy Corridor: A Reality', speech delivered at the Washington Business Forum, 25 February. http://nationalinterest.org/article/from-the-caspian-to-the-mediterranean-the-east-west-energy-corridor-is-becoming-2260.

Crandall, M.S. (2006) *Energy, Economics, and Politics in the Caspian Region: Dreams and Realities*. Westport, CT: Praeger Security International.

Cross, J.C. (1969) *The Economics of Bargaining*. New York: Basic Books.

Currie, J.M, J.A. Murphy, and A. Schmitz (1971) 'The Concept of Economic Surplus and Its Use in Economic Analysis', *Economic Journal* 81(324), December, pp. 741–799.

Dodsworth, J.R. et al. (2002) 'Cross-border Issues in Energy Trade in the CIS Countries', IMF Policy Discussion Paper PDP/02/13, Washington DC.

ECOWAS (Economic Community of West African States) (n.d.). http://www.ecowas.int/.

Eden, L. et al. (2004) 'From the Obsolescing Bargain to the Political Bargaining Model', Bush School Working Paper 403, Texas A&M University, College Station, TX.

EIA (Energy Information Administration) (n.d.). http://www.eia.gov.

EIA (Energy Information Administration) (n.d.) 'International Dry Natural Gas Consumption Information'. http://www.eia.gov/pub/international/iealf/table13.xls.

Emerson, R. (1969) 'Operant Psychology and Exchange Theory', in R. Burgess and D. Bushell (eds), *Behavioural Sociology*, pp. 378–405. New York: Columbia University Press.

Emerson, R.M. (1976) 'Social Exchange Theory', *Annual Review of Sociology* 2, pp. 335–362.

Energy Charter Secretariat (n.d.). http://www.encharter.org.

Energy Charter Secretariat (1996) *The Energy Charter Treaty and Related Documents: A Legal Framework for International Energy Cooperation*. Brussels: Synergy Programme/EU. http://www.encharter.org/fileadmin/user_upload/document/EN.pdf.

ESMAP (Energy Sector Management Assistance Programme) (2003) 'Cross-border Oil and Gas Pipelines: Problems and Prospects', ESMAP Technical Paper 035, UNDP/World Bank/ESMAP.

Esteban, J. and J. Sakovics (2003) 'Endogenous Bargaining Power', Working Paper 644, Department of Economics and Business, Universitat Pompeu Fabra, Barcelona.

European Information Service (2003) 'EU/Russia: Further Setback for Energy Charter Negotiations', European Information Service, 1 July. http://www.europolitics.info/eu-russia-further-setback-for-energy-charter-negotiations-artr185466-10.html.

ExxonMobil (n.d.). http://www.exxonmobil.com.

ExxonMobil (2002) 'Chad/Cameroon Development Project', Report No. 7, Section 7, 2nd Quarter. Updated version at http://www.esso.com/Chad-English/PA/Files/8_allchapters.pdf.

Foster, C.D. (1992) *Privatization, Public Ownership and the Regulation of Natural Monopoly*. Oxford: Blackwell.

GhanaWeb (2004) 'Nigeria Grants Ghana US$40 Million Loan', July. http://www.ghanaweb.com/GhanaHomePage/NewsArchive/artikel.php?ID=62904.

Heath, A. (1976) *Rational Choice and Social Exchange: A Critique of Exchange Theory*. Cambridge: Cambridge University Press.

Homans, G.C. (1958) 'Social Behaviour as Exchange', *American Journal of Sociology* 63(6), pp. 597–606.

Hubert, F. and S. Ikonnikova, 2003 'Investment Options and Bargaining Power in the Eurasian Supply Chain for Natural Gas', paper prepared for the BIEE Conference, Oxford,

IDA/IFC (International Development Association/International Finance Corporation) (2005) 'Country Partnership Strategy for Georgia', World Bank paper, 12 October.

IEA (International Energy Agency) (2002) 'Background Papers on Security of Supply', Workshop on Security of Gas Supply, IEA Headquarters, Paris, June.

Ignotus, M. (pseud. for Henry Kissinger) (1975) 'Seizing Arab Oil: How the US Can Break the Oil Cartel's Stranglehold on the World', *Harper's Magazine*, March. http://www.thefez.net/etc/articles/Harpers_SeizingArabOil_Kissinger.pdf.

Ikonnikova, S. (2006) 'Games the Parties of Eurasian Gas Supply Network Play: Analysis of Strategic Investment, Hold-up and Multinational Bargaining', discussion paper, first draft. http://www.necsi.edu/events/iccs6/viewpaper.php?id=277.

Intergovernmental Agreement among The Republic of Turkey, The Azerbaijan Republic and Georgia relating to the transportation of petroleum via the Baku Tbilisi Ceyhan Main Export Pipeline, 1999.

IFC (International Finance Corporation) (n.d.). http://www.ifc.org.

IFC (International Finance Corporation) (n.d.) 'IFC Investments in Georgia'. http://www.ifc.org/ifcext/eca.nsf/Content/Georgia_Investment%20Table.

IFC (International Finance Corporation) (n.d.) 'IFC Investments in Turkey', http://www.ifc.org/ifcext/eca.nsf/Content/Turkey_investment%20table.

IFC (International Finance Corporation) (2003) 'Economic, Social and Environmental Overview of the Southern Caspian Oil and Gas Projects', World Bank briefing paper, February.

Janz, T. (2000) 'The Evolution and Diversity of Relationships in Canadian Families', research paper, Law Commission of Canada. http://www.samesexmarriage.ca/docs/family_evolution.pdf.

Johnston, D. (1994) *International Petroleum Fiscal Systems and Production Sharing Contracts*. Tulsa, OK: Pennwell Books.

Kargin V. (2003) 'Uncertainty of the Shapley Value'. New York: Cornerstone Research.

Keirstead, B.S. and D.H. Coore (1946) 'Dynamic Theory of Rents', *Canadian Journal of Economics and Political Science* 12(2), pp. 168–172.

Kemper, R. (2003a) 'End Game', comments on prospects for finalisation of the Energy Charter Transit Protocol, Energy Charter Secretariat, Brussels.

Kemper, R. (2003b) 'New International Rules on Energy Transit Close to Finalisation', *Russian Petroleum Investor*, March, pp. 28–31.

Kennedy, J.L. (1993) *Oil and Gas Pipeline Fundamentals*. Tulsa, OK: Pennwell Books.

Koskela, E. and R. Stenbacka (2000) 'Compensation and Bargaining with Entrepreneurship as the Outside Option', research paper, Department of Economics, University of Helsinki, Helsinki, May.

Koutsoyiannis, A. (1979) *Modern Microeconomics*. London: Macmillan.

Krueger, A.O. (2005) 'Macroeconomic Policies for EU Accession', speech delivered at the Central Bank of Turkey Conference, Ankara, Turkey, 6 May.

Lawler, E.J. (2001) 'An Affect Theory of Social Exchange', *American Journal of Sociology* 107, pp. 321–352.

Layard, P.R.G. and A.A. Walters (1978) *Micro-Economic Theory*. London: McGraw-Hill.

Liesen, R. (1999) 'Transit under the 1994 Energy Charter Treaty', *Journal of Energy and Natural Resources Law* 17(1), pp. 56–73.

Lindblom, C.E. (1948) 'Bargaining Power in Price and Wage Determination', *Quarterly Journal of Economics* 62(3), pp. 396–417.

Mansley, M. (2003) 'The Baku-Tbilisi-Ceyhan Pipeline and BP: A Financial Analysis (Building Tomorrow's Crisis?)'. London: Platform/Claros Consulting.

Martin, S. (1994) *Industrial Economics: Economic Analysis and Public Policy*. Englewood Cliffs, NJ: Prentice Hall.

Masseron J. (1990) *Petroleum Economics*. Paris: Editions Technip.

McLellan, B. (1992) 'Transporting Oil and Gas – the Background to the Economics', *Oil and Gas Finance and Accounting* 7(2).

Michaels, J.W. and J.A. Wiggins (1976) 'Effects of Mutual Dependency and Dependency Asymmetry on Social Exchange', *Sociometry* 39(4), December, pp. 368–376.

MEES (1997) 'US Opposed to Any Pipeline Crossing Iran', *Middle East Economic Survey* 40(43), October.

MEES (1998) 'AIOC Head Says Baku-Ceyhan Could Lose Money', *Middle East Economic Survey* 41(48), November.

MEES (2004) 'Finance Package for BTC Pipeline Project Signed in Baku', *Middle East Economic Survey* 47(6), February.

Morningstar, R. (2003) 'From Pipe Dream to Pipeline: The Realisation of the Baku-Tbilisi-Ceyhan Pipeline', speech delivered at Harvard Kennedy School of Government, Cambridge, MA, 8 May.

Muthoo, A. (1999) *Bargaining Theory with Applications*. Cambridge: Cambridge University Press.

Nash, J. (1953) 'Two Person Cooperative Games', *Econometrica* 21, pp. 128–140.

Ndum, F.N. (2004) 'The Chad/Cameroon Oil and Pipeline Project: What Does Cameroon Stand to Gain or Lose?', unpublished dissertation, University of Dundee.

Noreng, O. (2000) 'The Great Caspian Pipeline Game', *Internasjonal Politikk* 58(2), pp. 161–164.

Omonbude, E.J. (2002) 'How Feasible Is a West African Market for Natural Gas?', unpublished dissertation, Centre for Energy, Petroleum and Mineral Law and Policy, University of Dundee.

Omonbude, E.J. (2007) 'The Transit Oil and Gas Pipeline and the Role of Bargaining: A Non-technical Discussion', *Energy Policy* 35, pp. 6188–6194.

Omonbude, E.J. (2009) 'The Economics of Transit Oil and Gas Pipelines: A Review of the Fundamentals', *OPEC Review* 33(2), pp. 125–139.

Osborne, M.J. and A. Rubinstein (1990) *Bargaining and Markets*. San Diego, CA: Academic Press.

Pen, J. (1952) 'A General Theory of Bargaining', *American Economic Review* 42(1), p. 24.

Petroleum Economist (2005) 'Backing Baku', October. http://www.petroleum-economist.com.

Pipeline Industries Guild (1984) *Pipelines: Design, Construction and Operation*. New York: Longman.

Pulse of Turkey (1998) 'Energy – through Great Cooperation, Not the "Great Game"', 13 May. http://www.turkpulse.com/energy.htm.

Roberts, J. (2004) 'The Turkish Gate: Energy Transit and Security Issues', *Turkish Policy Quarterly* 3(4), p. 17.

Roth, A.E. (1978) 'The Nash Solution and the Utility of Bargaining', *Econometrica* 46, pp. 587–594.

Roth, A.E. (1979) 'An Impossibility Result concerning N-Person Bargaining Games', *International Journal of Game Theory* 8(3), September, pp. 129–132.

Schelling, T.C. (1960) *The Strategy of Conflict*. Cambridge, MA: Harvard University Press.

SCP Project ESIA (2002) 'Project Alternatives', final report. http://www.bp.com/liveassets/bp_internet/bp_caspian/bp_caspian_en/

STAGING/local_assets/downloads_pdfs/s/scp/esia/georgia/azspu_hsse_pmt_00246_a9.pdf.

Sen, A. (2000) 'Multidimensional Bargaining under Asymmetric Information', *International Economic Review* 41(2), May, pp. 425–450.

Sharkey, W.W. (1982) *The Theory of Natural Monopoly*. Cambridge: Cambridge University Press.

Socor, V. (2002) 'Russia Unties Its End of Caspian Knot', *Wall Street Journal Europe*, 27–29 September.

Soligo, R. and A. Jaffe (1998) 'The Economics of Pipeline Routes: The Conundrum of Oil Exports from the Caspian Basin', working paper, Baker Institute for Public Policy, Rice University, Houston, TX, April.

Stauffer, T.R. (2000) 'Caspian Fantasy: The Economics of Political Pipelines', *Brown Journal of World Affairs* 7(2), pp. 63–78.

Stern, J. (2006) 'The Russian-Ukrainian Gas Crisis of January 2006', research paper, Oxford Institute for Energy Studies.

Stern, J. (2009) 'The January 2009 Russia-Ukraine Crisis and the Imperative of Bypass Pipelines', *Baltic Rim Economies* 27(2), p. 18.

Stevens, P. (1996) 'A History of Transit Pipelines in the Middle East: Lessons for the Future', CEPMLP Seminar Paper SP23, University of Dundee.

Stevens, P. (2000a) 'Future Oil Prices: Influences and Instability', *CEPMLP Internet Journal* 8(13). http://www.dundee.ac.uk/cepmlp/journal/html/vol8/article8-13.html.

Stevens, P. (2000b) 'Pipelines or Pipedreams? Lessons from the History of Arab Transit Pipelines', *Middle East Journal* 54(2), pp. 224–241.

Stevens, P. (2009) 'Transit Troubles: Pipelines as a Source of Conflict', Royal Institute of International Affairs, London.

Stevenson, D. (2004) 'The Potential of the Chad/Cameroon Pipeline Project to Promote Sustainable Economic Development', research paper, Pennsylvania State University, University Park, PA. http://forms.gradsch.psu.edu/diversity/sroppapers/2004/StevensonDamon.pdf.

Thibaut, J.W. and H.H. Kelley (1959) *The Social Psychology of Groups*. New York: Wiley.

Tilton, J.E. (2004) 'Determining the Optimal Tax on Mining', *Natural Resources Forum* 28, pp. 144–149.

Tonge, D. (2002) 'Oil, Power and Gas: Near East Perspectives', paper presented at the CWC Associates Conference, Geneva, 8 July.

Van Gelder, J.W. (2004) 'The Financing of the Baku-Tbilisi-Ceyhan Project', research paper prepared for Focus on Finance.

Vernon, R. (1971) *Sovereignty at Bay: The Multinational Spread of U.S. Enterprises.* New York: Basic Books.

Vincent-Genod, J. (1984) *Fundamentals of Pipeline Engineering.* Houston, TX: Gulf Publishing.

Vinogradov, S. (2001) *Cross-border Oil and Gas Pipelines: International Legal and Regulatory Regimes,* research paper prepared for the Association of International Petroleum Negotiators.

Waelde, T.W. (ed.) (1996) *The Energy Charter Treaty: An East-West Gateway for Investment and Trade,* International Energy and Resources Law and Policy Series 10, July. London: Kluwer Law International.

Waelde, T.W. (2004) 'Legal and Policy Implications of a Relationship of Two International Treaties in Natural Energy Resources: OPEC and the ECT', *Oil, Gas & Energy Law* 5, December.

Waern, K.P. (2002) 'Transit Provisions of the ECT and the Energy Charter Protocol on Transit', *Journal of Energy and Natural Resources Law* 20(2), p. 172.

Weir, F. (2004) 'Oil Pipeline Sparks Controversy in Poor Georgian Village', *Alexander's Gas and Oil Connections.* http://www.gasandoil.com/news/2004/03/ntr41095.

Winrow, G.M. (2004) 'Turkey and the East-West Gas Transportation Corridor', *Turkish Studies* 5(2), pp. 23–42.

Worcester, D.A. (1946) 'A Reconsideration of the Theory of Rent', *American Economic Review* 36(3), June, pp. 258–277.

World Bank (n.d.). http://www.worldbank.org.

World Bank (n.d.) 'Project Overview: Chad-Cameroon Petroleum Development and Pipeline Project'. http://go.worldbank.org/LNOXOH2W50.

Yedidiah, S. (1980) *Centrifugal Pump Problems: Causes and Cures.* Tulsa, OK: Pennwell Books.

Yergin, D. (1991) *The Prize: The Epic Quest for Oil, Money, and Power.* New York: Simon and Schuster.

Index

African Union, 132
Amerada Hess, 61
Armenia, Shah Deniz gas field/
    SCP project, transit
    route value, 81, 82
Azerbaijan
    BTC oil pipeline project,
      bargaining power of, 89
    BTC oil pipeline project,
      cost responsibilities
      of, 73
    economic and political
      conditions of, 83
    relevance of East-West
      Energy corridor to,
      68-69
    Shah Deniz gas field/
      SCP project, cost
      responsibilities of, 74
Azerbaijan International
    Operating Company, 80

Baku-Novorossiysk oil
    pipeline, see Northern
    Route Export Pipeline
Baku-Supsa oil pipeline, see
    Western Route Export
    Pipeline
Baku-Tbilisi-Ceyhan
    (BTC) oil pipeline
    project
    bargaining positions of
      participants in, 88-89
    commitment tactics, 97
    economic and geo-political
      relevance to transit
      country, 76-77
    overview of, 60-62
    political motives for
      development of, 66-69
    project costs of, 72-74
    strategic investments and,
      128
    transit route value, 79-80
    transit route value and
      bargaining power,
      96-97
Baku-Tbilisi-Ceyhan Pipeline
    Company (BTC Co.),
    72, 73
bargaining power
    counter-balancing, 99
    of participants, economic
      and political factors
      influencing, 88-92,
      139-140
    of transit country,
      influencing factors,
      93-94
bargaining power outcomes,
    36-37, 39, 50-51
    influencing factors, 51-54
bargaining principles, 38-50
    implications for case studies,
      92-100
    and outcomes, 51-54
bargaining theory, 7-8, 55-56
    literature review, 36-38

behaviour of transit country, *see* transit country behaviour
benefit-sharing, 20–21
Benin
    economic conditions of, 86
    FDI role in, 129
    importance to international visibility, 132
    strategic alliances of, 134
    WAGP project, bargaining power of, 91–92
    WAGP project development, motives of, 70
    WAGP project, economic relevance to, 78
    WAGP project, transit route value of, 82
BP plc, 61, 62, 130
BTC oil pipeline project, *see* Baku–Tbilisi–Ceyhan oil pipeline project
bygones rule, 4, 22, 27

Cameroon
    Chad–Cameroon oil pipeline project, bargaining power of, 90
    Chad–Cameroon oil pipeline project development, motives of, 71
    Chad–Cameroon oil pipeline project, economic relevance to, 78–79
    Chad–Cameroon oil pipeline project, transit route value of, 82–83
    economic and political conditions of, 87–88
    FDI role in, 129
    strategic alliances of, 134
Cameroon Oil Transportation Company (COTCO), 75, 78–79
capital costs, influencing factors, 13–14
case studies
    comprehensive mutual dependency solution, 140–141
    implications of bargaining principles, 92–100
    project costs, 72
    project development, political and commercial motives, 65–66
    selection criterion of, 58
    see also Baku–Tbilisi–Ceyhan oil pipeline project; Chad–Cameroon oil pipeline project; Shah Deniz gas field/SCP project; West African gas pipeline project
Caspian oil energy
    export routes for, 80, 81–82
    strategic importance of, 67–69
Chad
    Chad–Cameroon oil pipeline project, bargaining power of, 90
    Chad–Cameroon oil pipeline project development, motives of, 71
    economic and political conditions of, 87
Chad–Cameroon oil pipeline project
    bargaining positions of participants in, 88–89
    economic and geo-political relevance to transit country, 78–79
    overview of, 65
    political and commercial motives for development of, 71–72
    project costs of, 75
    strategic investments and, 129
    transit route value, 82–83
    transit route value and bargaining power, 97
Chanturia, Giorgi, 131–132
ChevronTexaco Corporation, 64, 65, 71
commercially driven projects, 66
    Chad–Cameroon oil pipeline development, 71–72
commitment tactics
    bargaining and, 47–49
    cost of revoking, impact on bargaining power, 97–98
common trade interest
    bargaining implications of, 92–93

as bargaining prerequisite, 40
impact on bargaining outcome, 52
Communauté Electrique du Bénin, 75
competition for markets, impact on transit country behaviour, 120–121
complementary projects, impact on transit country behaviour, 130–131, 135
connected national pipelines model, 20
ConocoPhillips Company, 61
corruption, 83–84
cost-sharing, impact on bargaining outcomes, 54
COTCO, see Cameroon Oil Transportation Company
cross-border and transit oil and gas pipelines, 9
  bargaining theory and, 39–50
  economics of, 19–21
  factors influencing success/failure of, 58–60
  obsolescing bargain, 4, 5
  problems of, 3
cross-border and transit oil and gas pipeline agreements
  ECTP provisions on scope of, 111, 139
  types of, 20
cross-border trade in oil and gas, 2–3

disruption of pipelines, see pipeline disruptions
domestic energy markets
  of Georgia, 130–131
  pipeline project relevance to, impact on bargaining outcome, 54

East-West Energy corridor, 68–69, 133–134
EBRD, see European Bank for Reconstruction and Development
economic benefits, influence on transit country bargaining power, 93
Economic Community of West African States (ECOWAS), 64, 86
  dependence of Benin and Togo on, 132–133
economic rent, 23–24
  transit fees and, 25–32, 33
economics of pipelines, see pipeline economics
economy/economic environment
  of Azerbaijan, 83–84
  of Benin, 86
  of Cameroon, 87–88
  of Chad, 87
  of Georgia, 84–85
  of Ghana, 86–87
  of Nigeria, 85
  of Togo, 86
  of Turkey, 84
ECOWAS, see Economic Community of West African States
ECT, see Energy Charter Treaty
ECTP, see Energy Charter Transit Protocol
Elf Aquitaine SA, 62
endogenous mutual dependency, 123, 135–136, 140, 142
Energy Charter Protocol on Energy Efficiency and Related Environmental Aspects, 107
Energy Charter Secretariat, 103, 105, 106, 112, 114, 141
Energy Charter Transit Protocol (ECTP), 6, 102
  origins and objectives of, 105–106, 110
  provisions of, 110–112
  role in problem resolution, 112–115, 141
Energy Charter Treaty (ECT), 6, 102
  origins and objectives of, 103–105
  provisions of, 106–110
  role in problem resolution, 112–115, 141
energy efficiency, ECT provisions on, 107–108
energy in transit, ECTP provisions on, 111

Index

energy markets, types and size of, impact on pipeline economics, 20
Energy Sector Management Assistance Programme (ESMAP), 9
energy trade, ECT provisions on, 107
energy transport facilities, 109–110
Eni S.p.A., 61
environmental policy, impact on pipeline economics, 21
equity participation, *see* ownership/equity participation
ESMAP, *see* Energy Sector Management Assistance Programme
European Bank for Reconstruction and Development (EBRD), 130
  financing of BTC oil pipeline project, 72
  loan to SOCAR, 74
European Energy Charter, 103, 104
exogenous mutual dependency, 123, 135, 136, 144
ExxonMobil, 65

fair treatment, 110, 113–114
foreign direct investment (FDI)
  impact on Georgia's behaviour, 127, 128
  impact on transit country behaviour, 121
  impact on Turkey's behaviour, 127–128
  by transit country, 143
freedom of transit, 108–109
Friends of the Earth Ghana, 70

game-theoretic bargaining solutions, 7–8, 37
gas pipeline projects
  transit country bargaining power, influencing factors, 90–92
  versus oil pipelines, bargaining outcomes of, 54–55
  versus oil pipelines, technical requirements of, 15–16

gas transmission costs, components of, 16
General Agreement on Tariffs and Trade (GATT), 6, 102, 107, 115
  Article V, 112
Georgia
  BTC oil pipeline project, bargaining power of, 89
  BTC oil pipeline project, cost responsibilities of, 73–74
  BTC oil pipeline project, economic relevance to, 76
  BTC oil pipeline project, transit route value of, 80
  economic and political conditions of, 83
  FDI role in, 127, 128
  importance to international visibility, 131–132
  role in East–West Energy corridor, 68–69
  Shah Deniz gas field/SCP project, bargaining power of, 91
  Shah Deniz gas field/SCP project, cost responsibilities of, 74
  Shah Deniz gas field/SCP project, economic relevance to, 77–78
  Shah Deniz gas field/SCP project, transit route value of, 81, 82
Ghana
  economic and political conditions of, 86–87
  WAGP project development, motives of, 70–71
  WAGP project, Nigerian financial aid to, 75

IFC, *see* International Finance Corporation
IMF, *see* International Monetary Fund
information asymmetry, bargaining and, 49–50
Inpex Corporation, 61
inside and outside options

Index

impact on bargaining outcomes, 44–47
impact on transit country bargaining power, 96–97
International Finance Corporation (IFC), 60
  assistance to Benin and Togo, 129
  assistance to Cameroon, 129
  assistance to Georgia, 127
  assistance to Turkey, 128
  financing of BTC oil pipeline project, 72
  financing of Chad–Cameroon oil pipeline project, 71–72, 75
International Monetary Fund (IMF), 83, 94
  assistance to Georgia, 84
international organisations, 6
  transit country bargaining power and, 94
  see also International Finance Corporation; International Monetary Fund; World Bank
international relations
  transit country bargaining power and, 95
  transit country behaviour and, 131
  transit country behaviour and strategic alliances, 133–134
interrelated agreements, 20
investment protection and promotion
  ECT provisions on, 106–107
  see also strategic investments
IPI gas pipeline project, see Iran–Pakistan–India gas pipeline project
Iran
  BTC oil pipeline project, transit route value of, 80
  Shah Deniz gas field/SCP project, transit route value of, 81–82
Iran–Pakistan–India (IPI) gas pipeline project, 126–127, 135
Itochu Corporation, 61

liquefied natural gas (LNG), 3, 19, 81, 82
  growth in cross-border trade of, 2
  load factor, 15, 22
LukAgip N.V., 62

military force, bargaining power and, 47
mutual dependency
  as comprehensive solution to transit fee disputes, 140–141, 142–143
  features of, 123–134, 136
  notion of, 121–123
  types of, 123

Nash bargaining solution, 37
natural monopoly, 18–19
Nigeria
  economic and political conditions of, 85–86
  financial assistance to Ghana, 75
Nigerian National Petroleum Corporation, 64
Nord Stream pipeline development, 66
Northern Route Export Pipeline (NREP), 60–61, 80

obsolescing bargain
  cross-border oil and gas pipelines, 4, 5
  notion of, 3–4
  producer country, 46
off-take transit country, 90–91
  bargaining outcome and, 52–53
  Benin and Togo as, 78, 91–92
  commitment tactics and bargaining power of, 49
  dependence on off-take, 120
  Georgia as, 91
  inside/outside options and bargaining power of, 46
  risk aversion and bargaining power of, 43
  time factor and bargaining power of, 41–42

oil pipeline projects
  bargaining power of transit country, influencing factors, 89–90
  versus gas pipelines, bargaining outcomes of, 54–55
  versus gas pipelines, technical requirements of, 15–16
oil price volatility, 7
operating costs, 14
operating pipeline bargaining situations, 51
ownership/equity participation
  in BTC oil pipeline project, 61–62, 72–73
  in Chad–Cameroon oil pipeline project, 65, 75
  influence on transit country bargaining power, 93–94
  in Shah Deniz gas field/SCP project, 62–63, 74
  in WAGP project, 64, 75

partial commitments, 47–49
  cost of revoking, 97–98
patience
  bargaining and, 40–42
  transit country bargaining power and, 94
Petronas, 65
pipelines
  alternative, impact on bargaining outcomes, 53
  alternative, impact on transit country bargaining power, 96–97
  alternative, impact on transit country behaviour, 119–120
  complementary, impact on transit country behaviour, 130–131, 135
  as natural monopolies, 18–19
pipeline economics, 11–13
  of cross-border and transit pipelines, 22–32
  of cross-border pipelines, 19–21
  of cross-border pipelines, implications, 21–22
  technical factors of feasibility of, 14–17
pipeline disruptions, 7
  resolution role of ECT and ECTP in, 112–115, 144
pipeline size, impact on pipeline economics, 12, 13–14, 16–17
political/diplomatic relations
  impact on bargaining outcomes, 54
  impact on cross-border and transit pipeline economics, 21
  impact on risk aversion in bargaining, 44
politically driven projects, 66
  BTC oil pipeline development, 66–69
  Shah Deniz gas field/SCP development, 69
  WAGP development, 70–71
politics/political environment
  of Azerbaijan, 84
  of Cameroon, 88
  of Chad, 87
  of Georgia, 85
  of Ghana, 87
  impact on transit country behaviour, 120
  of Nigeria, 86
  of Turkey, 85
pre-pipeline bargaining situations, 50–51, 56
producer countries
  commitment tactics and bargaining power of, 48–49
  inside/outside options and bargaining power of, 46–47
  risk aversion and bargaining power of, 43–44
  time factor and bargaining power of, 42
  see also Azerbaijan; Chad; Nigeria
producer surplus, 24
project costs, 28–32
  of BTC oil pipeline project, 72–74

of Chad–Cameroon oil pipeline
project, 75
risk aversion and, 44
of Shah Deniz gas field/SCP project, 74
of WAGP project, 74–75
project value, 28
pumping costs, 15
pure economic rent, 23
pure transit country
commitment tactics and bargaining power of, 49
Georgia as, 76
inside/outside options and bargaining power of, 46
risk aversion and bargaining power of, 43
time factor and bargaining power of, 41
transit fees importance and bargaining outcome, 53

quasi-rents, 25

regulation/regulatory regimes, 19–20, 102
see also Energy Charter Transit Protocol; Energy Charter Treaty
risk aversion
bargaining and, 42–44
transit country bargaining power and, 94–96
Russia, 21, 114
BTC oil pipeline project, opposition from, 67
BTC oil pipeline project, transit route value of, 80, 95
Nord Stream pipeline development, 66
Shah Deniz gas field/SCP project, transit route value of, 81, 95
strategic investments by, 125
Turkey as competitor to, 68

Shah Deniz gas field/South Caucasus Pipeline (SCP) project
bargaining positions of the participants in, 91
economic and geo-political relevance to transit country, 77–78
overview of, 62–63
political motives for development of, 69
project costs of, 74
strategic investments and, 127–128
transit route value, 80–82
transit route value and bargaining power, 96
Shapley value, 8, 137n. 2
Shell, 64
SOCAR, see State Oil Company of the Azerbaijan Republic
Société Beninoise de Gaz, 64, 93–94
Société Togolaise de Gaz, 64, 93–94
South Caucasus Pipeline (SCP), see Shah Deniz gas field/South Caucasus Pipeline project
State Oil Company of the Azerbaijan Republic (SOCAR), 61, 62
EBRD loan to, 74
stake in BTC Company, 73
Statoil ASA, 61, 62
strategic alliances
impact on transit country behaviour, 133–134
strategic investments
bargaining power and, 124–130
subadditivity, 18

time factor
bargaining power and, 41–42
transit country bargaining power and, 94
Togo
economic conditions of, 86
importance to international visibility, 132–133
FDI role in, 129
strategic alliances of, 134

160 Index

Togo – *continued*
  WAGP project, bargaining power of, 91–92
  WAGP project development, motives of, 70
  WAGP project, economic relevance to, 78
  WAGP project, transit route value of, 82
Total S.A., 61
transfer earnings, 24–25, 27
transit
  ECT notion of, 108, 109
  ECT provisions on, 108–110
transit country
  bargaining power of, influencing factors, 89–94
  compensation to, *see* off-take transit country; transit fees; transit tariffs
  FDI by, 141
  protection of rights of, 113
  role of strategic investments in, 124–130
  see also Benin; Cameroon; Georgia; Togo; Turkey
transit country behaviour, 4–5, 7
  influencing factors, 119–121
  see also mutual dependency
transit fees, 5–6, 32–33
  agreements, 58
  for Cameroon, 78–79
  determining factors, 33
  dispute solution, *see* mutual dependency
  ECTP provisions on, 112
  for Georgia, 77–78
  importance to transit country, 53, 120
  notion of, 25–26
  purpose of, 11
  purpose and justification of, 26–32
  see also transit tariffs
transit oil and gas pipelines, *see* cross-border and transit oil and gas pipelines

transit route value
  BTC oil pipeline project, 79–80
  Chad–Cameroon oil pipeline project, 82–83
  impact on transit country bargaining power, 96–97
  Shah Deniz gas field/SCP project, 80–82
  WAGP project, 82
transit tariffs
  ECTP provisions on, 111–112
  see also transit fees
transmission costs, 16
Turkey
  BTC oil pipeline project, bargaining power of, 89–90
  BTC oil pipeline project, cost responsibilities of, 74
  BTC oil pipeline project, economic and geo-political relevance to, 76–77
  BTC oil pipeline project, transit route value of, 80
  economic and political conditions of, 85
  FDI role in, 127–128
  role in East–West Energy corridor, 68, 69
  Shah Deniz gas field/SCP project, cost responsibilities of, 74
  Shah Deniz gas field/SCP project, transit route value of, 81–82
Turkiye Petrolleri Anonim Ortakligi, 61, 62, 74

United States
  strategic interests in BTC oil pipeline project, 67–69
  strategic interests in Georgia, 85
Unocal Corporation, 61
utilisation factor, 15

value
  of a project, 28
  of transit route, 79–83, 96–97
Vernon, Raymond, 3

DOI: 10.1057/9781137274526

Volta River Authority (Ghana), 64, 74–75

WAGP, *see* West African gas pipeline project
WAPCo, *see* West African Gas Pipeline Company
West African Gas Pipeline Company (WAPCo), 74–75, 93–94
West African gas pipeline (WAGP) project
   bargaining power of the participants in, 91–92
   economic and geo-political relevance to transit country, 78
   overview of, 63–64
   political and commercial motives for development of, 70–71
   project costs of, 74–75
   strategic investments and, 128
   transit route value, 82
Western Route Export Pipeline (WREP), 61
   value of, 80
World Bank, 60, 70, 78, 83, 115
   assistance to Benin and Togo, 129
   assistance to Chad, 87
   assistance to Georgia, 84
   role in Chad–Cameroon oil pipeline development, 71, 75
   withdrawal of support to Chad–Cameroon oil pipeline project, 65, 71–72
WREP, *see* Western Route Export Pipeline

Lightning Source UK Ltd.
Milton Keynes UK
UKOW040211301012

201376UK00001B/23/P